NEW CHALLENGES IN SEED BIOLOGY - BASIC AND TRANSLATIONAL RESEARCH DRIVING SEED TECHNOLOGY

Edited by **Susana Araújo**
and **Alma Balestrazzi**

New Challenges in Seed Biology - Basic and Translational Research Driving Seed Technology

http://dx.doi.org/10.5772/61583
Edited by Susana Araújo and Alma Balestrazzi

Contributors

Irfan Afzal, Zubaida Yousaf, Trinidad Ruiz Téllez, José Blanco Salas, Javier Carbonell Espín, Francisco María Vázquez Pardo, Byoung Ryong Jeong, Prabhakaran Soundararajan, Abinaya Manivannan, Mehmet Cetin, Hakan Sevik, Kerim Guney, Kudret Betül Güney, Stanley Lutts, Paolo Benincasa, Lukasz Wojtyla, Szymon Kubala, Katzzarina Lechowska, Muriel Quinet, Malgorzata Garnczarska, Alicia Gamboa-De Buen, Ximena Gómez-Maqueo, Mikio Nishimura, Masatake Kanai, Shoji Mano, Makoto Hayashi

CBS, Edition **2017**

Published by InTech

Janeza Trdine 9, 51000 Rijeka, Croatia

Publishing Process Manager Andrea Koric

Technical Editor SPi Global

Cover Designer InTech Design Team

Additional hard copies can be obtained from orders@intechopen.com

New Challenges in Seed Biology - Basic and Translational Research Driving Seed Technology
Edited by Susana Araújo and Alma Balestrazzi
p. cm.
Print ISBN 978-953-51-2658-4
Online ISBN 978-953-51-2659-1

INTECH

open science | open minds

Contents

Preface

Seed biology, a fascinating aspect of plant sciences, covers all the crucial aspects of seed development, germination, as well as seed treatments. The need for increased seed quality has become a priority necessary to face the current demand for high standards in the agricultural market and seed industry [1, 2]. Seed producers are always looking for high-quality seeds to fit the demands of the competitive EU seed market, which reached a value around € 7 billion, corresponding to 20 % of the global market in 2012 [3].

In this book *New Challenges in Seed Biology - Basic and Translational Research Driving Seed Technology*, we provide a state of the art of the current knowledge on seed biology with focus on seed technology. A collection of eight chapters written by seed biology experts from the field of seed physiology, ecology, molecular biology, biochemistry, and industrial seed technology was gathered.

Optimal crop performance requires rapid and uniform seedling emergence since delayed germination exposes plantlets to unfavorable environmental conditions. Critical points in the search for novel hallmarks of seed vigor are the metabolic changes induced by water uptake and the response to priming or invigoration treatments. In this context, Lutts and coworkers contributed with an updated review on the molecular mechanisms underlying the beneficial response of seeds to priming, taking advances of the recent knowledge gathered by the use of global expression profiling methods. In the same topic, Afzal looks over the recent advances on seed treatments, highlighting the use of physical treatments as alternative approaches to chemical-based seed invigoration protocols.

Knowledge on the metabolic dynamics that accompany the transition from the quiescent state of dry seeds to the active proliferating state of germinating seeds/seedlings is being continuously updated. The signaling cross talk between the reactive oxygen species and growth regulators implicated in the release and/or induction of dormancy was discussed by Soundararajan et al. Mehmet et al. described the effects of the application of plant regulators on seed germination of *Lilium* genus, highlighting the potential of these types of seed treatments for mass propagation of both endangered and ornamental *Lilium* species.

Seed storage plays a major role in ensuring seed quality from the moment the seeds reach physiological maturity until they germinate. Appropriate postharvest handling helps to preserve seed viability but also ensures the economical value of the final product. In developing countries, seed postharvest techniques are a proxy of increased food security and income for local populations. Yousaf et al.reviewed the different postharvesting techniques available to address the needs of not only developing countries but also the more industrialized ones.

Advanced molecular tools applied to translational research programs in seed science are expected to address key societal challenges in agriculture and industry while ensuring environmental protection. Kanai et al. extensively described how the available fundamental knowledge on lipid biosynthesis could be used for increasing oil production in crops. Gómez-Maqueo et al. highlighted the potentialities of modulating cell wall properties to enhance seed germination traits. On the other side, Ruiz-Téllez et al.reviewed the potentialities of seed technology for helophyte production to be used in green wetlands with horizontal surface for wastewater treatment.

We hope that this book, which combines different aspects of basic and translational research in seed biology, will attract the attention of researchers and technologists from academia and industry, providing points for interactive and fruitful discussion on this fascinating topic.

Acknowledgments

This work was supported by grants from the University of Pavia. During this book edition, SSA has been awarded by a research contract funded by CARIPLO Foundation (Action 3, Code 2013-1727)—Integrated Project "Advanced Priming Technologies for the Lombardy Agro-Seed Industry (PRIMTECH)." The financial support from Fundação para a Ciência e a Tecnologia (Lisbon, Portugal) is acknowledged through research unit "GREEN-it: Bioresources for Sustainability" (UID/Multi/04551/2013) and presents SSA postdoctoral grant (SFRH/BPD/108032/2015).

References

[1] Araújo SS, Paparella S, Dondi D, Bentivoglio A, Carbonera D, Balestrazzi A. Physical methods for seed invigoration: advantages and challenges in seed technology. Frontiers in Plant Science 2016 7: 646. DOI: 10.3389/fpls.2016.00646

[2] Paparella S, Araújo SS, Rossi G, Wijayasinghe M, Carbonera D, Balestrazzi A. Seed priming: state of the art and new perspectives. Plant Cell Reports 2015 34: 1281–1293. DOI: 10.1007/s00299-015-1784-y

[3] Ragonnaud M. The EU seed and plant reproductive material market in perspective: a focus on companies and market shares (2013). Policy Department B: Structural and Cohesion Policies. European Parliament Committee on Agriculture and Rural Development. Brussels: European Commission

Dr. Susana Araújo
Plant Cell Biotechnology Laboratory
Oeiras, Portugal

Alma Balestrazzi
Universitá degli Study di Pavia,
Pavia, Italy

Chapter 1

Seed Priming: New Comprehensive Approaches for an Old Empirical Technique

Stanley Lutts, Paolo Benincasa, Lukasz Wojtyla,
Szymon Kubala S, Roberta Pace,
Katzarina Lechowska, Muriel Quinet and
Malgorzata Garnczarska

Additional information is available at the end of the chapter

http://dx.doi.org/10.5772/64420

Abstract

Seed priming is a pre-sowing treatment which leads to a physiological state that enables seed to germinate more efficiently. The majority of seed treatments are based on seed imbibition allowing the seeds to go through the first reversible stage of germination but do not allow radical protrusion through the seed coat. Seeds keeping their desiccation tolerance are then dehydrated and can be stored until final sowing. During subsequent germination, primed seeds exhibit a faster and more synchronized germination and young seedlings are often more vigorous and resistant to abiotic stresses than seedlings obtained from unprimed seeds. Priming often involves soaking seed in predetermined amounts of water or limitation of the imbibition time. The imbibition rate could be somehow controlled by osmotic agents such as PEG and referred as osmopriming. Halopriming implies the use of specific salts while "hormopriming" relies on the use of plant growth regulators. Some physical treatments (UV, cold or heat,..) also provide germination improvement thus suggesting that priming effects are not necessarily related to seed imbibition. A better understanding of the metabolic events taking place during the priming treatment and the subsequent germination should help to use this simple and cheap technology in a more efficient way.

Keywords: germination, omics approaches, priming, seedling growth, stress resistance

INTECH
open science | open minds

1. Introduction

Efficient seed germination is important for agriculture. Successful establishment of early seedling indeed requires a rapid and uniform emergence and root growth. Germination of orthodox seeds commonly implies three distinct phases (**Figure 1**) consisting in (1) Phase I: seed hydration process related to passive imbibition of dry tissues associated with water movement first occurring in the apoplastic spaces; (2) Phase II: activation phase associated with the re-establishment of metabolic activities and repairing processes at the cell level; and (3) Phase III: initiation of growing processes associated to cell elongation and leading to radicle protrusion. Phases I and III both involve an increase in the water content while hydration remains stable during Phase II. It is commonly considered that before the end of Phase II, germination remains a reversible process: the seeds may be dried again and remain alive during storage and able to subsequently re-initiate germination under favorable conditions.

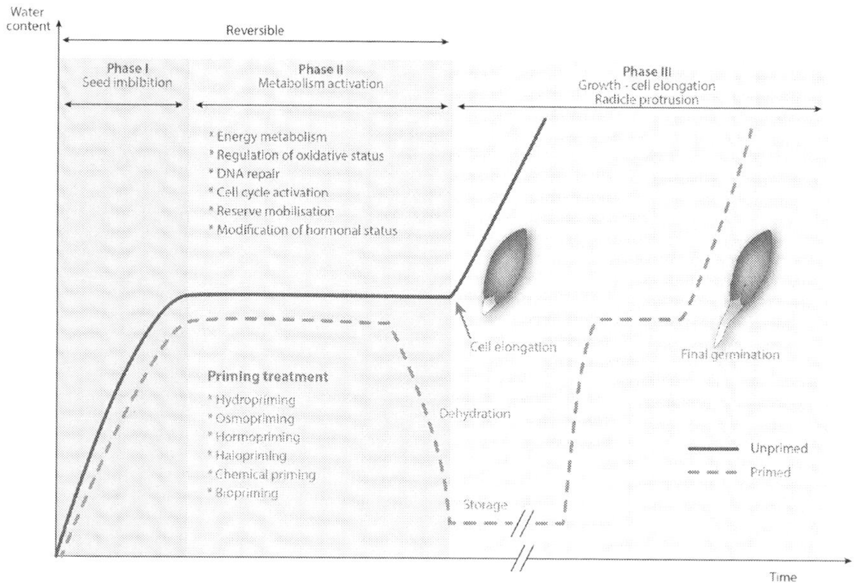

Figure 1. Seed hydration curves and germinating phases in unprimed and primed seeds.

Water-based seed priming is defined as a pre-sowing treatment that partially hydrates seeds without allowing emergence [1]. Various treatments may indeed be applied during the reversible phase of germination (point 3). They widely differ according to the osmotic potential of the priming solution, the duration, the external temperature, and the presence of specific chemical compounds. The efficient treatments trigger metabolic processes activated during the phase II of germination, which are then temporally stopped before a loss of desiccation occurs (**Figure 1**) [2].

The overall consequence of seed priming consists in an increased seed vigor defined as the whole set of properties conditioning seed lots performance in a wide range of environment [3]. Priming strategies may afford several economic and agronomic advantages to cultivated plants (point 4). Numerous data published in the literature indeed reported an improvement in the rate and uniformity of germination but also an obvious improvement in the behavior of the obtained seedlings in terms of plant growth and stress resistance.

Although priming is used since decades by farmers and seed companies to improve germination, it can also occur under natural plant conditions. This is more especially the case in serotinous plants growing in deserts and able to retain their seeds for a long time. These seeds indeed undergo several hydration-dehydration cycles enhancing subsequent germination after final seed dispersion caused by heavy rain [4]. From a general point of view, the priming process not only concerns seeds but also the whole plant system itself and may be defined as an induced state whereby a plant reacts more rapidly and more efficiently to a stress [5]. In this acception, plants exposed to a primary constraint are triggering a set of temporary metabolic adaptation leading to a stress memory and allowing them to adapt more efficiently to subsequent episodes of stress [6, 7].

Although the interest of seed priming has been demonstrated since a long time, the underlying physiological and biochemical basis of this fascinating process remain poorly understood. Holistic approaches related to omics tools now provide new opportunities to elucidate the molecular components of the priming phenomena. Similarly, nondestructive and noninvasive methods such as digital image technology may be used in a more precise way to study the kinetics of imbibition in relation to the modification of the seed ultrastructure. This chapter reviews the most recent progresses accomplished in the understanding of the seed priming-induced modifications.

2. A brief history of seed priming

Man established contact with seed physiology since the beginning of agriculture and quickly realized that many seeds do not germinate easily and uniformly. Ancient civilization was fascinated by the capacity of an apparently « dead seed » to resurrect and to produce a viable young and healthy seedling after germination. The Greek Theophrastus (ca. 372–287 BC) already focused on seed physiology and suggested that germination process may be temporarily interrupted [8]. Pre-hydration of legume seeds before sowing was performed by Roman farmers in order to increase the germination rate and synchronize germination as reported by the Roman naturalist Gaius Plinius Secundus. Several centuries later, these techniques were still used for a wide range of species according to the French agronomist Olivier de Serres (1539–1619) [8]. In 1664, Evelyn [9] mentioned that temperature prior sowing may have an impact on further germination while one century later, Ingenhousz [10] analyzed the impact of light on seedling emergence.

During the nineteenth century, numerous botanists started to describe morphological processes associated with seed germination [11, 12]. Sachs [13] experimented the impact of various compounds (including tyrosine and asparagine) before and during germination. The discovery of plant hormones in the 1920s underlined the crucial role of these compounds in seed desiccation tolerance, reserve mobilization, as well as cell division and cell elongation occurring during germination. The possibility to influence final germination as a consequence of pre-sowing treatment has led to a wide range of empirical methods for numerous cultivated plant species during the year 1970s [14].

3. Priming methods and priming agents

Several methods of seed priming have been developed in order to invigorate seeds and alleviate the environmental stresses. A common feature of water-based priming techniques, which distinguishes them from other pre-sowing treatments, is partial seed pre-hydration and the activation of early germination events in seed. Priming efficiency is affected by many factors and strongly depends on treated plant species and chosen priming technique. Physical and chemical factors such as osmotica and water potential, priming agent, duration, temperature, presence or absence of light, aeration, and seed condition also influence priming success and determine germination rate and time, seedling vigor, and further plant development [15, 16].

3.1. Hydropriming

Hydropriming is the simplest method of seed priming, which relies on seed soaking in pure water and re-drying to original moisture content prior to sowing. No use of additional chemical substances as a priming agent makes this method a low-cost and environmentally friendly. The main disadvantage of hydropriming is uncontrolled water uptake by seeds. This is a consequence of free water availability to seeds during hydropriming, so that the rate of water uptake depends only on seed tissue affinity to water [17]. Moreover, this technique may result in unequal degree of seeds hydration thus leading to lack of simultaneous metabolic activation within seeds followed by unsynchronized emergence [18]. Considering these limiting factors, it is highly important to define accurate treatment duration, temperature, and water volume used in hydropriming to ensure desired level of seed hydration and to prevent radicle protrusion. Despite the aforementioned limitations, many reports indicated beneficial effect of hydropriming on seed germination and seedling growth under both optimal and stress conditions, in various crop plants such as chickpea, maize [19], wheat [20], Indian mustard [21], canola [22], sunflower [23], rice [24], mung bean [25], capsicum [26], and durum wheat [27].

One of the commercially used types of hydropriming is the system named "drum priming", patented in the early 1990s [28, 29]. In this technique, seeds are gently rotating in drum and gradually hydrated by addition of water in vapor form. Drum priming allows seed imbibition in a controlled manner and could be an attractive alternative to conventional hydropriming. Specially designed apparatus enables monitoring of the seed weight, precise regulation of time, and water amount during hydration process, what ultimately results in an appropriate and uniform moisture level of the seeds [30]. Drum priming with 24-epibrassinolide shows positive effect on germination time and seedling growth of bell pepper concomitant with improved superoxide dismutase (SOD), catalase (CAT), and peroxidase (POX) activities [31]. Another variant of hydropriming, so-called "on-farm priming", consist of seed soaking in water followed by surface drying and subsequent sowing. The duration of treatment obligatorily cannot be longer than "safe limit" (maximum time of priming without risk of seed or seedling damage by premature germination) [32]. The positive impact of this method on crop emergence and yield was confirmed by Harris et al. [33]. On-farm priming is especially useful for resource-poor farmers in marginal tropical environment [34].

3.2. Osmopriming

Osmopriming involves soaking seeds in osmotic solution with low water potential instead of pure water. Due to low water potential of osmotic solutions, water enters seed slowly which allows gradual seed imbibition and activation of early phases of germination but prevents radicle protrusion [35]. Usually water potential of priming agent varies from −1.0 down to −2.0 MPa [36]. However, values of water potential together with duration of the priming treatment should be always adjusted to species, cultivar, and sometimes seed lot. Different compounds are used in osmopriming procedure including polyethylene glycol (PEG), mannitol, sorbitol, glycerol, and inorganic salts such as NaCl, KCl, KNO_3, K_3PO_4, KH_2PO_4, $MgSO_4$, and $CaCl_2$ [37]. Priming with salt solutions is often referred as "halopriming". Most common chemical employed in osmopriming treatment is PEG, mainly owing to its specific characteristic. Large molecular size of PEG prevents its penetration into the seed thus avoiding induction of potential cytotoxic effect and reduction of osmotic potential within seed [35]. Nevertheless, PEG exhibits some undesirable features including high viscosity, which restrict diffusion of oxygen in the solution so in PEG priming aeration system is preferred [2]. Seed priming with PEG has been shown as an effective method to improve seed germination, seedling emergence, and stress tolerance of several crop plants under unfavorable conditions such as salt, water, chilling, and nano-ZnO stresses [1, 3, 36–38].

3.3. Solid matrix priming

Solid matrix priming (SMP, matriconditioning), in which water uptake by seeds is controlled, has been developed as an alternative method to osmopriming because of high cost of osmotic agents and technical problems with aeration [2]. During solid matrix priming, seeds are mixed and incubated with wet solid water carrier for a certain period. Afterward, seeds are separated from matrix, rinsed, and back-dried. The use of solid medium allows seeds to hydrate slowly and simulates natural imbibition process occurring in the soil [18]. To successfully accomplish SMP, materials utilized as matrices should possess specific physical and chemical features such as low matrix potential, minimal water solubility, high water holding capacity and surface area, no toxicity to seeds, and ability to adhere to seed surface. In fact, vermiculite, peat moss, charcoal, sand, clay, and some commercially offered substrate such as Celie or Micro Cell are exemplary solid carries applied in solid matrix priming [2, 35]. In order to obtain the best priming performance, time of treatment and optimal water content must be determined separately for each matrix [39].

Positive effects of SMP on crop seeds have been noted in many reports. Solid matrix priming enhanced field performance of carrot [40] as well as improved germination and seed vigor of soybean [41]. Study on onion showed that matriconditioning improved seed germination rate, seedling emergence, and growth under optimal and low temperature conditions [42]. Sand priming increased the activities of antioxidant enzymes such as catalase (CAT), peroxidase (POX), and soluble sugar content in waxy maize concomitant with improved rate of germination and seedling growth under high-salt stress conditions [43].

It is well established that integration of SMP with biological and chemical factors may greatly enhance seed performance [18]. Adoreoli and de Adnrade [44] indicated that inclusion of gibberellins/fungicide/*Bacillus subtilis* to matriconditioning leads to improved stand establishment and productivity of some vegetable crops under tropical conditions. Similarly, matriconditioning with GA_3 enhanced the quality of hot pepper seeds [45]. More recently published data demonstrated that solid matrix priming with *Trichoderma viride* improved seedling emergence and yield of okra under low temperatures [46].

3.4. Hormopriming

During hormopriming, seeds imbibition occurs in the presence of plant growth regulators, which can have direct impact on seed metabolism. The following regulators are commonly used for hormopriming: abscisic acid, auxins, gibberellins, kinetin, ethylene, polyamines, and salicylic acid (SA). Gibberellic acid (GA_3) and PEG priming improved photosynthetic properties, antioxidant system, seedling emergence, and growth of white clover on heavy metal polluted soil [47]. Priming spring wheat seeds with GA_3 increased grain yield and salt tolerance by modulating hormone homeostasis together with alterations of ion uptake and accumulation between shoots and roots [48]. Enhanced salt tolerance, growth, and grain yield of wheat were also observed after kinetin-priming [49]. Among the different techniques of seed priming (hydro-, osmo-, and halopriming), spermidine pretreatment appeared to be the most effective method for induction drought tolerance in rice [50]. High efficiency of polyamines-priming on the improvement of rice tolerance to drought has been demonstrated also by Farooq et al. [51]. Critical role of phytohormones exogenously supplied into seeds for plant response to salinity stress was stated in wheat seeds primed with ascorbic acid and salicylic acid, as this pretreatment method increases the ability of wheat to grow successfully under salt stress, whereas hormonal priming with ABA was not effective in this case [52].

3.5. Biopriming

Biopriming involves seed imbibition together with bacterial inoculation of seed [53]. As other priming method, this treatment increases rate and uniformity of germination, but additionally protects seeds against the soil and seed-borne pathogens. Hydration of seeds infected with pathogens during priming can result in a stronger microbial growth and consequently impairment of plant health. However, applying antagonistic microorganisms during priming is an ecological approach to overcome this problem [54]. Moreover, some bacteria used as biocontrol agents are able to colonize rhizosphere and support plant in both direct and indirect way after germination stage [55]. It was found that biopriming is a much more effective approach to disease management than other techniques such as pelleting and film coating [56]. Nowadays, the use of biopriming with plant growth-promoting bacteria (PGPB) as an integral component of agricultural practice shows great promise [57, 58]. In pearl millet, biopriming with *Pseudomonas fluorescens* isolates enhanced plant growth and resistance against downy mildew disease [59]. Biopriming with rhizobacteria improved germination parameters of radish seeds under saline conditions [60].

3.6. Others

Chemical priming refers to seed treatment with different chemical solutions used as priming agents. This approach includes priming with wide range of both natural and synthetic compounds such as antioxidants (ascorbic acid, glutathione, tocopherol, melatonin, and proline), hydrogen peroxide, sodium nitroprusside, urea, thiourea, mannose, selenium, chitosan, fungicide etc. Positive impact of chemical priming with various priming agents in a wide range of environmental conditions was indicated by numerous studies [26, 61–64]. Seed priming with β-amino butyric acid increased drought and salt tolerance of green gram [65]. Application of ascorbic acid as a seed priming agent induced drought and salt resistance of wheat [66, 67]. Analysis conducted by Fercha et al. [67] revealed that priming with ascorbate counteracts the negative effects of salinity stress by changes in abundance of proteins involved in metabolism, protein destination, and storage.

Nutripriming is a technique in which seeds are soaked with solutions containing the limiting nutrient instead of pure water. The idea of this method is to obtain nutritional effect together with biochemical advantages of priming in order to improve seed quality, germination parameters, and seedling establishment [68]. Seed priming with Zn improved productivity of chickpea and wheat [69], germination and early seedling growth of rice [70], development and root growth of maize seedling exposed to low root zone temperatures [71], while K-priming brought favorable effect on growth and nutrient status of cotton seedling under saline conditions [72]. Some nutripriming techniques are commonly used by seed companies in the process of seed production and preparation for growers. One of this methods, broad spectrum nutrient seed priming (BSN), is based on imbibing seeds in mixture of minerals, such as zinc, copper, manganese, molybdenum, and phosphorus, which has been proved to fertilize the seed and provides the nutrients for early growth, which positively affects germination, seedling vigor, and root system development (http://seedprimer.com/).

4. Seed priming and agriculture

Pre-sowing priming induces a particular physiological status in seeds and has emerged as a promising strategy to improve plant behavior in the field. There is a strong interest for farmers and seed companies to find suitable cheap priming treatments but also to precisely identify the agronomical properties improved as a result of priming in cultivated species.

4.1. Hastening and synchronization of germination

Primed seeds often exhibit an increased germination rate and greater germination uniformity. An enhanced and uniform seedling emergence may contribute to regular crop establishment. Priming may enhance events taking place at the beginning of the germination, but the whole process is interrupted at a given state, which is the same for all concerned seeds. Priming may also induce structural and ultrastructural modifications that could facilitate subsequent water uptake and attenuate initial differences between the seeds in terms of imbibition, thus resulting in a more uniform germination [47].

A faster emergence may help to improve competitivity of cultivated plants against weed species as recently demonstrated by Jalali and Salehi [73] for sugar beets. In mung bean plants, a faster seedling establishment resulting from priming may contribute to a total increase in yield up to 45% [74].

Priming-induced increase in germination may be associated to a change in plant hormone biosynthesis and signaling. Priming has been reported to increase gibberellins (GA)/abscisic acid (ABA) ratio [75], and this may be a direct consequence of a priming impact in gene expression pattern [76]. A more uniform GA endogenous concentration in primed seeds may help to synchronize endosperm weakening, embryo cell elongation, and reserve mobilization [77]. Ethylene also directly influences germination speed and percentage. Increase in ethylene production during priming may promote endo-β-mannase activity facilitating endosperm weakening and post-priming germination [78]. Priming has been reported to initiate repair and reactivation of pre-existing mitochondria and to initiate the biogenesis of new ones [79]. It may thus afford a higher level of energy over a short time to sustain final germination [80].

4.2. Plant growth

Plants issued from primed seeds often exhibit a faster growth than those issued from unprimed ones. Determine whether such growth stimulation is the consequence of a more rapid seedling establishment or result from a long-term specific physiological status induced by priming still remains an unresolved question. In numerous cases, the beneficial impact of priming on plant growth is more obvious under nonoptimal than under optimal conditions, leading to the global concept that a major advantage of priming consists in an increase in stress resistance (point 4.10). Thus, in direct relation to memory events, the main question is related to the remanence of priming-induced modifications. Imram et al. [71] showed that such modifications remain intact several weeks after germination in maize.

In rice, priming with 5-aminolevulinic acid improved shoot elongation [81] while priming with picomolar rutin augmented both root and shoot length in relation to an increase in photosynthetic pigments, phenolic and flavonoid contents [82]. In wheat, priming with sodium prusside stimulated plant growth as a consequence of improved capacity to scavenge free radicals by antioxidants [83], and a similar observation was reported for rice as a result of an increase in glutathione peroxidase (GPX) activity [24] and other antioxidant enzyme activities [84].

The beneficial impact of priming on plant growth may be due to an improved nutrient use efficiency allowing a higher relative growth rate [85] and to an improved regulation of the plant water status [86]. Jisha and Puthur [65] confirmed that the priming effect of β-aminobutyric acid on seeds of *Vigna radiata* further get carried over the seedlings. A higher growth of seedlings issued from primed seeds may also be analyzed in relation to a direct impact of pretreatment on cell cycle regulation and cell elongation processes (point 7) [77, 78].

4.3. Mineral nutrition

Modification of nutrient uses efficiency by young seedlings may be a consequence of priming-induced overexpression of genes encoding for specific transporters, although only few transporters appear specifically induced by priming itself [36]. An efficient strategy to improve mineral nutrition of young seedling is to use nutrient-based seed priming strategy. Phosphorous seed priming supported crop development at early stages and may compensate for P deficiency in the soil [87, 88]. Jamil et al. [89] demonstrated that improvement of mineral status of P-primed cereals reduced strigolactone exudation and thus sensitivity to the parasite weed *Striga hermonthica*. Muhammad et al. [85] recently performed experiments using Zn, Mn, B, and P priming. These authors demonstrated that nutrient seed priming allowed maize plants to maintain Zn and Mn supply for at least 3 weeks in highly calcareous soils characterized by a low nutrient availability. Similarly, Pame et al. [90] showed that P accumulation in rice may be increased by using P-primed seeds, which is of special interest in Asia where about one-third of the area of rainfed rice is situated on P-deficient soils. Such a higher absorption could not be explained only by nutrient accumulation in the seeds during the primed phase since it is still observed in plants several weeks after sowing. It may therefore be hypothesized that priming interferes with regulation of acquisition mechanisms and further research is crucially needed to identify the molecular mechanisms involved in these processes. Priming with boron improves seedling emergence in rice and, on a long-term basis, increases panicle fertility in relation to an improvement in stigma receptivity [91]. Seed priming may also contribute to improve N nutrition, mainly through an enhanced nitrate reductase activity in plants [40]. Priming with nonessential beneficial elements, such as Si, leads to an increase in Si content of cultivated plants and has a protective impact on plant development [86].

Beside the improvement of essential elements uptake, priming also helps to reduce accumulation of putatively toxic elements. Chromium (VI) accumulation is reduced in maize seedlings issued from salicylic acid primed seeds and cultivated in the presence of this toxic element [82]. Osmopriming with PEG and hormopriming with GA improved germination and early seedling growth of white clover maintained on a heavy metal-contaminated soil, but the impact on Cd accumulation by plants may differ according to the considered treatment since GA_3 increased Cd accumulation while PEG reduced it [47]. Liu et al. [92] demonstrated that PEG increases Ca^{2+} cytosolic concentration through hyperpolarization-activated calcium permeable channels, which could explain a lower Cd accumulation as a consequence of an improved selectivity toward calcium.

Numerous data are also available concerning the priming effect on the plant behavior exposed to salinity. It is frequently reported that priming-induced stress resistance may be a consequence of an improved discrimination for K^+ over Na^+ nutrition. Both osmo- and hydropriming were efficiently used to influence K^+ selectivity of seedlings, but the underlying molecular basis of this improvement still needs to be identified, especially in terms of regulation of monovalent cation transporters.

4.4. Yield-related parameters

A huge amount of studies is devoted to the impact of seed priming on the seed germination phase and early seedling growth. Most of those studies are conducted under controlled environmental conditions in plant growth chambers or greenhouses. Data reporting a real improvement under field conditions remain rare. Yield effect may be linked to a faster plant establishment allowing a longer growth period. Khan et al. [93] reported that plant issued from primed seed benefits from a longer period of assimilates accumulation in sugar beet. Conversely, in some cases, phenological evolution of cultivated plants may be modified by priming: in chickpea, plants issued from priming encountered an earlier seed maturity allowing them to escape disease or heat terminal stress in the season [94]. Yield increase may also result from a higher plant density observed as a consequence of priming-induced increase in germination percentage [95].

Since less than a decade, several data started to be available for priming-induced yield improvement in rice. Shah et al. [96] demonstrated that priming had a positive effect on the weight of 1000 grains in this species. Boron priming induced an obvious decrease in panicle sterility and consequently improved the number of grains per inflorescence [91]. Binang et al. [97] also demonstrated that priming had a significant effect on the number of tillers, number of fertile panicles, and consequently grain yield of new NERICA rice varieties. Promising yield improvement has also been reported for maize [85, 98], onion [99], okra [100], and sugar beet [73]. Beside its impact on quantitative parameters, priming may also improve the quality of harvested plants, as recently reported by Janecho et al. [101] for the vitamin content and nutritional value of legumes.

4.5. Stress resistance

Most of the studies performed on the seedlings issued from primed seeds demonstrated a clear improvement of resistance to environmental constraints. **Table 1** is providing a nonexhaustive list of recent publications dealing with stress resistance improvement on cultivated plant species. Frequently, such improvement is obvious just after emergence at the seedling level, but progressively disappears at the adult stage. For example, some young plants issued from priming treatments displayed improvement of resistance to chilling [84], low temperature [75], salinity [43, 102], high temperature [80], drought [24, 65, 103], and UV exposure [82]. Some interesting studies also demonstrated that priming may afford resistance to biotic stresses such as *Fusarium oxysporum* in tomato [104], viral disease in *Brassica rapa* [105], and downy mildew in pearl millet [106]. Such a large set of data suggests that seed priming may elicit numerous pathways contributing to stress resistance. The molecular basis involved in this stress resistance remains intact during the dehydration phase following priming and may contribute to stress resistance during the final germination step. Moreover, some data suggest that a single priming treatment may induce resistance to various stresses.

Environmental constraint	Plant species	Priming treatment	Reference
Salinity	*Brassica juncea*	hydro/osmopriming	[21]
	Brassica napus	PEG	[120]
		Halopriming	[22]
	Helianthus annuus	KNO_3	[23]
	Triticum durum	ascorbic acid	[27]
	Medicago sativa	PEG	[37]
	Zea mays	Sand priming	[43]
	Triticum aestivum	ABA + SA	[52]
		NO (nitroprusside)	[63]
		Ascorbic acid	[67]
		Biopriming	[102]
		KCl + $CaCl_2$	[167]
	Raphanus sativus	Biopriming	[60]
	Gossypium arvense	Potassium	[72]
	Citrus sinensis	NO (nitroprusside)	[185]
Drought/water stress	*Oryza sativa*	KH_2PO_4	[24]
		Dry priming	[50]
		Polyamines	[51]
		Salicylic acid	[134]
	Zea mays	Urea; KNO3	[64]
	Vigna radiata	BABA	[65]
	Triticim aestivum	Ascorbic acid	[66]
		Silicon	[86]
	Brassica napus	PEG	[178]
		Hydropriming	[117, 118]
	Glycine max	Osmoconditioning	[187]

Environmental constraint	Plant species	Priming treatment	Reference
	Cicer arietinum	Osmo/hydropriming	[170]
Chilling and low temperatures	*Vigna radiata*	Hydropriming/proline	[25]
	Zea mays	*chitosan*	[62]
		Nutrient priming	[71]
	Glycine max	Osmopriming	[79]
	Oryza sativa	Salicylic acid	[84]
	Beta vulgaris	Osmopriming	[93]
	Spinacea oleracea	Osmopriming	[107]
	Nicotiana tabacum	Putrescine	[166]
High temperatures	*Lactuca sativa*	Hydropriming	[76]
	Daucus carota	PEG	[80]
Heavy metals	*Trifolium repens*	PEG	[47, 179]
	Poa pratensis	PEG, Gibberellins	[179]
Biotic stresses			
Phythium ultimum	*Zea mays*	Biopriming	[53]
Verticillium	*Brassica napus*	Biopriming	[56]
Downy mildew	*Pennisetim glaucum*	Biopriming	[59]
		BABA	[106]
Yellow mosaic virus	*Vigna radiata*	On farm priming	[75]
Fusarium oxysporum	*Solanum lycopersicum*	Methyl jasmonate	[104]
Double-stranded DNA virus	*Brassica rapa*	Biopriming	[105]

Table 1. Nonexhaustive list of recent studies devotes to priming-induced increase in the stress resistance of cultivated plant species.

The priming procedure itself implies frequently the use of stressing agent, as it is the case for PEG and salt. In some cases, priming may be performed at low temperature to reduce the kinetics of seed hydration. A slow hydration may be considered as a stressing process since the water content is too low to allow radicle elongation (see point 5). It may thus induce defense responses within embryos. This is especially the case for biochemical processes involved in protection against reactive oxygen species (point 8). Several components of the ROS-mediated signaling pathways are activated during the first hydration phase of the priming process. The ultimate stress resistance in the seedlings may then be linked to the persistence of the antioxidative defenses after final germination. Since management of oxidative stress is an important component of resistance to a wide range of stress, this observation may, at least partly, explain the cross-resistance phenomena.

The dehydration step that follows the partial hydration phase is also a major stressing phase. Numerous studies focus on late embryogenesis abundant proteins (LEA) normally involved in the acquisition of desiccation tolerance. Chen et al. [107] showed that the major dehydrins disappear during osmopriming while Maia et al. [108] conversely suggested that PEG may induce LEA synthesis. Water deprivation associated with the dehydration phase may also trigger accumulation of transcription factors, some of them being specifically involved in stress resistance [109, 110]. Molecular chaperones such as heat shock proteins (HSP) also explain a priming-induced improvement of stress resistance [78]. It may also be hypothesized that priming-induced modification of the seed hormonal status, mainly an increase in ABA, may somewhat influence the seed and young seedling response to environmental constraints in relation to a faster activation of ABA-responsive genes involved in stress acclimation [108, 109].

The beneficial impact of priming treatments relies on numerous properties as indicated in **Figure 2**.

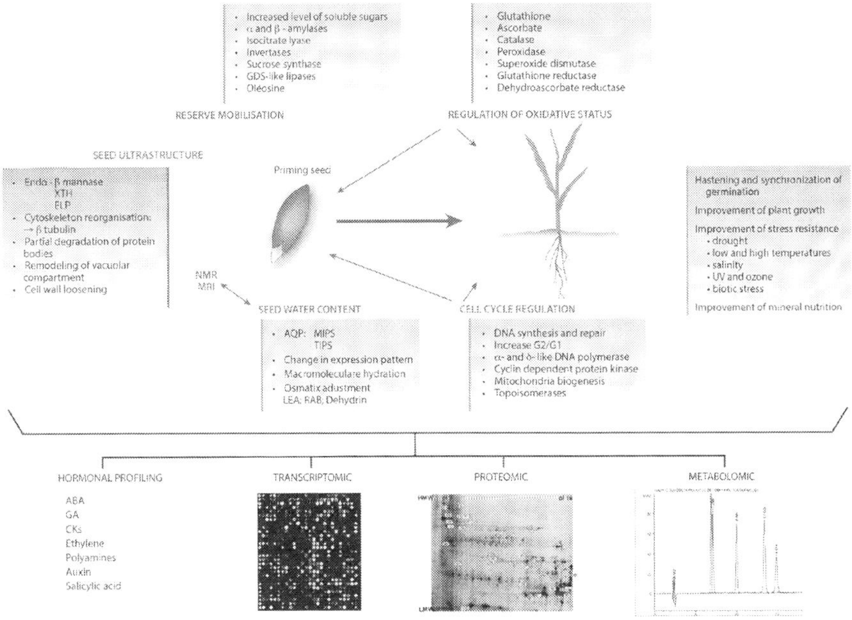

Figure 2. General overview of biochemical and physiological basis of priming effects. Priming modifies seed ultra-structure, reserve mobilization, regulation of oxidative status and cell cycle, and seed water content. The obtained seedlings may be improved for growth, mineral nutrition, and stress resistance. The components of priming effect may be revealed through an integrated convergent proteomic, transcriptomic, and metabolomics holistic approach.

5. Priming and seed water content

Seed germination is characterized by triphasic kinetic of water uptake with a rapid initial uptake (imbibition), followed by an apparent lag phase. A further increase in water uptake occurs only after germination is completed, as the embryonic axis elongates [111] (**Figures 1** and **2**). Early imbibition of seeds involves the fastest and most drastic changes in tissue hydration observed during germination. The water content of seeds and tissues within seeds depends of the composition of stored reserves. Seed imbibition and subsequent embryo growth depend on water exchanges and water potential gradients represent the motive force for water flow and finally tissue expansion. Water transport across cell membranes is essential for the initiation of metabolism. This intracellular water transport is mediated by aquaporins.

Aquaporins (AQPs) are transmembrane proteins, members of the major intrinsic protein (MIP) family that facilitate rapid and passive water transport across cell membranes and play a crucial role in plant water relations [112–114]. Plant aquaporins are remarkably diverse with several subfamilies of MIPs identified in dicots and monocots. Among them, the plasma membrane intrinsic proteins (PIPs) and the tonoplast intrinsic proteins (TIPs) subfamilies constitute the largest number of AQPs and correspond to AQPs that are abundantly expressed in the plasma and vacuolar membranes, respectively. Both PIP and TIP subfamilies are believed to play a key role in transcellular and intracellular plant water transport.

To gain insight into the role of water channel in germination, the expression profiles of *AQP* genes were studied in *Arabidopsis* [115], *Oryza sativa* [116], and *Brassica napus* [117] during seed imbibition and early embryo growth. These results have demonstrated the possible role of several AQPs in seed germination also in response to abiotic stresses. Moreover, Liu et al. [116] have shown the reduced seed germination rate via *OsPIP1;3* silencing and the promotion of seed germination via *OsPIP1;3* over-expression under drought conditions demonstrating that OsPIP1;3 is required for normal seed germination.

Seed priming involves imbibing seeds with restricted amounts of water to allow hydration sufficient to permit pre-germinative metabolic events to proceed while preventing radicle protrusion. This treatment can extend phase II of water uptake while preventing seeds from entering into phase III. The completion of radicle emergence is prevented by restricted amount of water provided to the seed (hydropriming, solid matrix priming) or decreased water potential (Ψw) of the imbibition medium by the use of osmotic solutes such as PEG or salts (osmopriming) [111]. In a study on *Brassica napus* osmopriming, Kubala et al. [36] revealed that at the beginning of the soaking period and at the end of drying phase, the seed water content was as low as 5%. The soaking treatment allowed seed imbibitions up to 50%, which should be enough to re-initiate metabolism.

As AQPs regulate water movement, it can be supposed that these proteins play an important role both in priming treatment (seed soaking) and post-priming germination under both favorable and unfavorable conditions. A role for aquaporin-controlled water transport across cell membranes in primed seeds of *Brassica napus* during germination was demonstrated by Gao et al. [118]. Seed priming with PEG or ABA resulted in an enhanced germination, particularly under salt and osmotic stresses at low temperature. Priming treatment induced expression of *BnPIP1* but had no effect on transcript level of Bnγ-*TIP2*. However, transcripts of both Bn*PIP1* and Bnγ-*TIP2* genes during germination were present earlier in primed seeds than nonprimed ones. Gao et al. [118] speculated that Bn*PIP1* was involved in water transport required for the activation of enzymatic metabolism of storage nutrients in the early stages of rapeseed germination, while Bnγ-*TIP2* expression was correlated with cell growth during radicle emergence. Changes in the expression pattern of *SoPIP1;1, SoPIP1;2, SoPIP2;1,* and *So*δ*TIP* during *Spinacia oleracea* seeds osmopriming and post-priming germination under optimal conditions, chilling and drought have been reported by Chen et al. [119]. The authors have stated that all these genes were up-regulated within 2–4 d of priming (phase II-imbibition). Therefore, the high expression of those *AQPs* might contribute to water transport across plasma and vacuolar membranes to facilitate water supply to expanding tissues and to increase germination potential of primed seeds. The down-regulation of all *AQP*s genes expression was observed under chilling and drought. However, the expression of some *AQPs* genes was elevated in primed seeds that also exhibited greater chilling and drought tolerance [119]. Kubala et al. [36] revealed up-regulation of two genes encoding tonoplast AQPs (*TIP4.1* and *TIP1.2*) in *Brassica napus* seeds in relation to osmopriming. In this study, expression of *TIP1.2* increased approximately 20 fold during post-priming germination as compared to unprimed seeds. In addition, the same authors have also stated facilitated water uptake and higher stress tolerance of germinating primed *Brassica napus* seeds [26, 120]. The above-mentioned results have demonstrated that water transport and sufficient water supply for embryo during post-priming germination regulated by AQPs may be one of the crucial components modulated by pre-sowing seed priming that influences germination rate and stress resistance.

As PIPs, but not TIP, are generally found at the plasma membrane, PIPs are thought to play a key role in seed water uptake. Nevertheless, both microarray [121] and macroarray experiments [115] with the complete set of genes encoding major intrinsic proteins revealed that out of 13 PIPs encoded by the *Arabidopsis* genome, transcripts for only three isoforms (PIP1;2, PIP1;4, and PIP1;5) were detectable in seeds. Mapping of TIPs in germinating *Arabidopsis* seeds has revealed that isoforms TIP3;1 and TIP3;2 detected in embryos, appear to localize to both the plasma membrane and tonoplast [122]. Vander Willigen et al. [115] have observed that during germination, very high level of TIP3 protein was coincident with decreased level of PIP1;2 and PIP2;1 polypeptides until phase III of water uptake. As stated by Vander Willigen et al. [115], it is intriguing how such low concentrations of PIP protein during the early phases of germination can achieve basic transcellular water transport in the seed. Gattolin et al. [122] have speculated that TIP3 may be the only AQP involved in seed water intake, and that the presence of TIP3 at the plasma membrane may compensate for the absence (or low concentration) of PIPs. In the light of these results, the enhanced germination potential of primed *Brassica napus* seeds could be partially explained by the up-regulation of TIPs during priming and post-priming germination [36]. However, the involvement of apoplastic water movement and simple diffusion of water across membranes during seed imbibition cannot be ruled out. The treatment of *Arabidopsis* seeds with mercury, a general blocker of aquaporins, reduced the speed of seed germination but did not affect its developmental sequence or basic aspects of seed water relations. Vander Willigen et al. [115] suggested that aquaporins functions are not involved in early seed imbibition but would rather be associated with water uptake accompanying expansion and growth of the embryo.

The transmembrane water transport via the regulation of AQP quantity and activity endows seeds with a remarkable capacity to modulate water absorption, transport, and compartmentation within tissues. Nuclear magnetic resonance (NMR) spectroscopy has provided insights into changes in the physical states of seed water during germination [123–126]. In particular, magnetic resonance imaging (MRI) has revealed a precise spatial distribution of water within tissues of germinating seeds and different patterns between species [125, 127, 128] highlighting the tight control of water transport. Water status of primed seeds was characterized by Nagarajan et al. [129] in study on tomato halo-and osmopriming. Nagarajan et al. [129] pointed out that better performance of primed seeds may be attributed to the modifications of seed water-binding properties and reorganization of seed water during imbibition, so as to increase the macromolecular hydration water required for various metabolic activities related to the germination process. In the future, it will be crucial to see how the spatial pattern of aquaporin expression can fit the hydration pattern revealed by MRI during both priming and post-priming germination, therefore enabling a comprehensive understanding of water transport in seeds.

Several studies have reported that water uptake is improved by priming treatment as primed seeds exhibited a faster imbibition in comparison with nonprimed ones, although pre-treated seeds were dried after priming to reach the same water content as nonprimed ones [36, 47, 107]. Although MRI studies revealed that water penetrates seeds through the hilum and micropyle [125, 130], Galhaut et al. [47] did not observe any particular modification of these structures after *Trifolium repens* priming, despite a faster seed hydration. However, scanning electron microscopy analysis showed that primed seeds of *Trifolium repens* exhibited seed coat tears and circular depressions that can favor seed imbibition. Moreover, X-ray photographs revealed tissue detachment in dry primed seeds that formed free space between the cotyledons and radicle, making water flow easier, thus contributing to tissue hydration [47]. Similarly, formation of free space surrounding the embryo in dry primed seeds of tomato was noticed by Liu et al. [131]. In brief, these observations suggest that structural modifications might contribute to rapid seed germination by improving water uptake.

The maintenance of favorable water status is critical for survival of germinating seeds under environmental stresses leading to tissue dehydration. The accumulation of nontoxic, compatible solutes within seed tissues, that is, osmotic adjustment is a major trait associated with maintenance of high cell turgor pressure potential in response to stress conditions. Priming treatment itself may generate a moderate abiotic stress during soaking (e.g. osmotic stress, salt, and drought created by the priming agents) [36, 132]. The accumulation of osmotically active solutes such as amino acids (e.g. proline) ammonium compounds (e.g. glycine betaine), sugars (e.g. glucose, fructose, sucrose) during priming was noticed in several species and was shown to improve seed germination under subsequent water stress [3, 21, 133, 134].

Seeds can also experience dehydration in the course of priming treatment, that is, during drying after soaking. Late embryogenesis abundant proteins (LEAs) can stabilize cell structure and macromolecules upon cell dehydration by preventing inactivation and aggregation of proteins and the loss of membranes integrity. This could be realized due to the ability of most LEA proteins to either coat intracellular macromolecules with a coherent layer of water or to interact with the surface of proteins and thus acting as water replacement [135]. As LEA proteins accumulate at a high level in response to cell/tissue dehydration, they may contribute to acquisition of tolerance to drought and related stresses such as osmotic, salt, and cold stress. In support of this, several studies revealed changes in the pattern of expression/accumulation of LEA transcript/protein in seeds caused by priming treatment and suggested their association with improved stress tolerance of primed seeds [36, 107, 136–138]. For example, the transcripts of two genes: *Em6*, encoding LEA group 1 protein and *RAB18*, encoding responsive to ABA 18 protein, belonging to LEA group 2, declined during osmopriming (soaking in PEG solution), reaccumulated after slow drying and again degraded during *Brassica oleracea* seeds germination [138]. The up-regulation of *RAB18* and *Em6* expression during slow seed drying suggests that they play a role in drought tolerance. Chen et al. [107] reported transient accumulation of four dehydrins-like proteins (32, 30, 26, 19-kD) in seeds of *Spinacia oleracea* during early stages of osmopriming followed by progressive degradation to a lower level in primed dry seeds compared to unprimed ones. A similar trend was confirmed to cold acclimation protein CAP85. In contrast to protein concentration, relative expression of *CAP85* was greater in primed dry seeds than in unprimed ones. Recently, Kubala et al. [36] revealed accumulation of LEA transcripts (*LEA4-1, LEA4-5*) and LEA3 proteins during soaking in PEG solution. The authors proposed that soaking in PEG with a low osmotic potential should not be considered only as a rehydration phase: water uptake may be sufficient to reinitiate a physiological activity from a previous quiescent stage but water content of 50% remained low enough to represent a water stress situation, especially when it is maintained during several days [36].

6. Priming and seed ultrastructure

In general, the ability of seeds to germinate seems to be critically determined by a change in the balance between growth potential of the embryo and the mechanical resistance of the surrounding tissues. In many species, the endosperm tissue enclosing the embryo restrains the germination process acting as a physical barrier, which restricts radicle emergence.

Weakening of tissues surrounding elongating radicle by cell separation due, for instance, to the activity of cell wall hydrolases may occur as a consequence of priming (**Figure 2**). It was determined that osmopriming induced hydrolysis of the endosperm tissue of *Cucumis melo* seeds [139] and increased the endo-β-mannanase activity in the endosperm cap and decreased its mechanical restraint on the elongating tomato embryo [140]. A strong correlation was observed between lowering of the mechanical restraint and the activity of endo-β-mannanase [141].

Penetration of the structures surrounding the embryo is a consequence of radicle cells elongation. Up-regulation of the gene encoding xyloglucan endotransglucosylase/hydrolase (XTH) in response to osmopriming and accumulation of transcript for extensin-like protein

(ELP) during post-priming germination was observed in rapeseed [36]. As XTHs have the ability to cleave xyloglucans and rejoin the cut ends with new partners, they are engaged in cell wall loosening during growth and in the restructuring of the cell walls after extension.

Cytoskeleton reorganization is also necessary to achieve large rates of cells elongation that precedes radicle protrusion. The component of microtubules (β-tubulin) accumulated in tomato seeds during germination and priming and the expression preceded visible germination [142]. Higher level of β-tubulin protein accumulation was shown in rapeseed during PEG soaking, drying, and post-priming germination. The up-regulation of genes encoding γ- and β-tubulins was also noticed during post-priming germination [36].

Ultrastructural observations performed during the 6-d period of solid matrix priming (SMP) of carrot (*Daucus carrota*) seeds indicated the breakdown of storage materials, specific to the catabolic phase of germination *sensu stricto*, both in the axis and in the micropylar endosperm covering the radicle tip [143]. It was found that complete degradation of storage protein and lipid bodies and subsequent starch accumulation occurred in the radicles of carrot seeds after 8-d SMP. In the endosperm, the catabolic changes were limited to the micropylar area, where extensive breakdown of storage cell walls, partial degradation of protein bodies, and no storage lipid hydrolysis were observed [144].

During seed germination, storage proteins, which provide a source of reduced nitrogen, and inorganic minerals need to be mobilized to support seedling growth. In addition, a lytic aqueous vacuolar compartment building up the turgescence necessary for cell expansion and to promote radicle protrusion and embryo elongation has to be formed (**Figure 2**). Bolte et al. [145] investigated the features and the dynamics of the vacuoles during the early stages of *Arabidopsis* seed germination and indicated the successive occurrence of two different lytic compartments in the protein storage vacuoles (PSV). The first one corresponds to globoids specialized in mineral storage and the second one is at the origin of the central lytic vacuole in these cells [145]. Different mechanisms for the transformation of PSV into lytic vacuole in the root tip cell of germinating tobacco (*Nicotiana tabacum*) seeds were proposed by Zheng and Staehelin [146]. Ultrastructural studies demonstrated that the radicle cells of tobacco contain only one type of vacuole at particular time of development. Upon rehydration, the radicle cells only contain PSVs, but during subsequent root development, the PSVs are systematically transformed into lytic vacuoles via cell type specific pathways.

At present, we do not have a complete view of ultrastructural changes occurring during seed priming. One would expect that similar vacuole remodeling might occur during priming, especially in embryonic axis. The maintenance of metabolism over, for instance, several days of seed priming requires mobilization of embryo reserves. It is speculated that accumulation of endogenous osmotica, cell vacuolization, together with cell wall loosening initiated during priming, might determine the embryonic axis extension and radicle protrusion during post-priming germination. Deeper and more detailed studies should be continued in order to completely clarify this phenomenon.

7. Seed priming and cell cycle regulation

Some of the hypotheses proposing explanation for priming-induced improvement are based on its effect on DNA in relation to activation of DNA repair mechanisms, synchronization of the cell cycle in G_2 and preparation to cell division (**Figure 2**). During seed maturation, most of the embryo cells are stopped at G_1 or G_0 phase of cell cycle and only some species have a small proportion of cells in the G_2 phase [111]. During seed imbibition, meristematic activity is limited; however, some preparation to cell division occurs. In embryos of dry tomato seeds, most cells have 2C DNA level and are in G1 phase of nuclear division [147]. The authors observed that DNA synthesis preceded germination as during imbibition in water, 4C signal was found mostly in the embryonic root tip, which suggests that cell enters S phase. They also primed seeds for 14 d in PEG-6000, which enhanced the rate and uniformity of germination. The 4C DNA signal of root tip cells increased during priming starting from 3 d incubation in PEG and was constant after re-drying the seeds to the initial moisture content. This observation suggests that priming increased the ratio of cells in G_2 phase to G_1 phase and indicates that the beneficial effects of priming on seedling performance are associated with the replicative DNA synthesis prior to germination [147]. This is accompanied by increase in α- and δ-like DNA polymerase activities in primed seeds and during germination.

The initiation of cell cycle and proceeding the cell to S phase may depend also on a G_1 checkpoint control. Most, if not all, cell cycle proteins responsible for cell cycle control appear to be already present in dry mature seeds, although some of them should be synthesized *de novo*. However, not only protein synthesis but also their modification may play a regulatory function for cell cycle control [148]. Cell division starts just after radicle protrusion, thus, seed priming, which prolongs Phase II of seed germination and is finished just before Phase III, does not affect cell division in itself [16]. Seed priming extends Phase II, when DNA repair mechanisms and expression of genes encoding proteins needed in cell cycle control and commencement are activated and overreach the level observed in unprimed seeds. Pre-activation of cell cycle by priming could be through regulation of the activity of cell cycle proteins such as cyclin-dependent protein kinases and proliferating-cell nuclear antigens [16]. It was found that osmopriming of *Brassica napus* seeds induced expression of cell division control protein 48 homolog C, cyclin P4;1, cyclin like protein and topoisomerase II in dry seeds, as well as proliferating-cell nuclear antigen 2 and cyclin dependent kinase 3;2 during imbibition [36]. Accumulation of proliferating-cell nuclear antigen during maize seed imbibition was associated with transition of the cells from G_1 to G_2 [149]. Moreover, microtubules, apart from cytoskeleton formation, cytoplasmic streaming, organellar movement, and cell wall formation, function in mitotic spindle formation during mitosis. Microtubules in dry seeds are depolymerized and form discrete granular bodies, which become organized into the cytoskeleton during imbibition [111, 142]. Higher expression of genes encoding microtubule-associated protein 65-1 and 70-2 as well as tubulin subunits γ-1, β-1, β-3 and microtubule motor activity proteins belonging to kinesin family was also observed during PEG soaking and in dry osmoprimed *Brassica napus* seeds [36]. Enhanced expression of tubulin genes was associated with accumulation of β-tubulin protein during osmopriming and subsequent germination [36]. Also in pre-hydrated seeds of *Arabidopsis thaliana* and *Solanum lycopersicum* accumulation of tubulins (mainly β-tubulin) was stated during germination as compared to unprimed seeds [136, 142].

During Phase II of seed germination, when water uptake is severely limited, major metabolic processes are activated [111]. One of the most important events undergoing during Phase II is DNA repair, which precedes cell cycle activation [142, 150]. The process of DNA replication is preceded by repair of DNA damage caused mainly by reactive oxygen species, which are accumulated during seed storage and aging [151]. DNA repair covers first period of DNA synthesis, while the second period of DNA synthesis (replication) is observed before cell division. DNA synthesis in Phase II of germination and also during seed priming corresponds rather to DNA repair, mainly in organelle such as plastids and mitochondria [152]. Increased number of mitochondria in leek embryo cell of osmoprimed seeds was observed by Ashraf and Bray [153]. Mitochondria biogenesis before mitochondria division involved the transition of promitochondria to mature mitochondria. This process is accompanied by the expression of genes of nucleotide biosynthesis, transport, and organelle RNA- and DNA-related functions [154, 155]. Pre-sowing seed osmopriming induced higher expression of genes corresponding to mitochondria biogenesis such as translocases of the inner membrane (TIM) complex TIM10 and TIM23-1, mitochondrial ribosomal protein and translational elongation factor EF2, which is targeted into mitochondria [36].

There are still some gaps in comprehensive understanding of pre-sowing seed priming impact on DNA repair and cell cycle regulation. Activation of different DNA repair mechanism has been observed during seed imbibition preceding germination and they are believed to be essential for successful reactivation of cell cycle [111]. They include α and β tyrosyl-DNA phosphodiesterase 1, α and β DNA topoisomerase I [156], 8-oxoguanine DNA glycosylase/lyase and formamidopyrimidine-DNA glycosylase [157], transcription elongation factor II-S [158], DNA ligase VI and IV [159]. Varier et al. [16] have suggested that in primed seeds DNA damage is repaired before replication, primarily through DNA synthesis. However, in a study on *Cicer arietinum* primed seeds, the role of DNA repair genes in enhancing the physiological quality of seeds was postulated [160]. The authors tested the expression level of genes encoding proteins with already proved function on DNA repair mechanisms in relation to priming methods and seed size. Moreover, enhanced accumulation of transcripts was found in dry and imbibed osmoprimed *Brassica napus* seeds [36] for genes involved in DNA repair according to function description in databases, such as DUTP-pyrophosphatase-like 1, endonuclease V family protein, ribonucleoside-diphosphate reductase subunit M2 (TSO), casein kinase II, replicon protein A2, DNA glycosylase DEMETER (DME), BARD1, RECQ helicase L4B, and MUTS homolog 2. Thus, activation of DNA repair mechanisms in seeds occurs prior to their germination and contributes to enhanced germination rate and better quality of seeds undergoing pre-sowing seed priming.

8. Management of oxidative status

Management of oxidative status is also considered as an important part of primed seeds physiology [18, 36, 161, 162] (**Figure 2**). Beginning with the seed development through maturation and germination, the seed moisture content as well as seed metabolic activity is subjected to dramatic changes. The biochemical and cellular events triggered by water uptake and subsequent loss are accompanied by a generation of reactive oxygen species (ROS) [151, 163]. During seed imbibition and early stages of germination, ROS production occurs mainly through respiratory activities of mitochondria, activities of β-oxidation pathways and enzymes such as NADPH oxidases, extracellular peroxidases, and oxalate oxidases [163]. ROS accumulation and associated oxidative damage can be regarded as a source of stress that may affect the successful completion of germination. As ROS, particularly H_2O_2 can act as signaling molecules, seeds must be endowed with a ROS removing system that tightly regulates their concentration. Scavenging of ROS is carried out by antioxidant system, a multifunctional machine, which includes enzymes (i.e. catalase (CAT), superoxide dismutase (SOD), and ascorbate peroxidase (APX), monodehydroascorbate reductase (DHAR), and glutathione reductase (GR)) as well as nonenzymatic compounds (i.e. ascorbic acid (AsA) or reduced glutathione (GSH) [1, 151]. Metabolism of ROS, mainly H_2O_2, which is believed to play a central role in oxidative status signaling, is strictly associated to other reactive species and signaling molecules such as nitric oxide and hydrogen sulfide, which contribute and regulate the transition from dormant to germination phase [151, 163, 164].

Kubala et al. [36] have considered the priming as a process, which consists of two main phases: controlled hydration of the seeds and drying back to the initial moisture content. Seed prehydration followed by re-drying during priming treatment, similarly to seed maturation and germination, exerts changes in moisture content, which leads to ROS production and activation of the antioxidant system. The activation of APX and the accumulation of AsA and GSH during osmopriming of spinach seeds (*Spinacia oleracea*) were accompanied by the repression of SOD and CAT activity [1]. These results indicated that activation of AsA-GSH cycle during osmopriming of spinach seeds can decrease the level of lipid peroxidation products in primed seeds. Moreover, the same authors have suggested that the renewal of antioxidant defense system, possibly required by seedling establishment, occurred during the late stages of germination as a result of up-regulation of CAT activity after initial reduction and overall antioxidant activation [1]. Kubala et al. [36] have indicated through an integrated transcriptomic and proteomic approach that the priming-induced germination can be linked with the activation of antioxidant system. The authors showed that during osmopriming of *Brassica napus* seeds *CAT2* and *PER21* encoding peroxidase 21 were up-regulated and GR protein was accumulated. Moreover, the same authors have observed up-regulation of *PER13* gene and accumulation of peroxidase 12, DHAR and peroxiredoxin proteins during post-priming germination. Furthermore, Kubala et al. [162] have postulated a correlation between activation of antioxidant metabolism in osmoprimed *Brassica napus* seeds and increased tolerance to salt stress during germination. The enhanced activity of APX, SOD, and CAT corresponded with increased expression rate of *APX, SOD,* and *CAT* genes. Similar result has been obtained by Nouman et al. [165], who showed that priming of *Moringa oleifera* seeds with *Moringa* leaf extract (MLE) improves growth under saline condition mainly by activation of antioxidant system (SOD, CAT, and POD). Priming of mung bean seeds (*Vigna radiata* L. Wilczek) with β-amino butyric acid (BABA) enhanced the activities of SOD and POX leading to improved tolerance to NaCl and PEG 6000 stresses [65]. Enhanced chilling stress tolerance in two tobacco varieties (MSk326 and HHDJY) was due to increased activity of antioxidant enzymes (SOD, POD, CAT, and APX) as a result of priming seeds with putrescine [165]. Results obtained by Islam et al. [167] showed that in haloprimed wheat (*Triticum aestivum*) seeds, increased activity of CAT, POD, and APX enhanced tolerance to salinity stress. Osmopriming with PEG has improved sorghum (*Sorghum bicolor*) seed germination and seedling establishment under adverse soil moisture conditions and has been correlated with antioxidant system activation (APX, CAT, POD, and SOD) [3]. Rice (*Oryza sativa*) seeds primed with polyethylene glycol (PEG) showed increased activity of APX in parallel with decreased activity of SOD, POD, and CAT under ZnO nanoparticles stress [38]. The same authors have also observed down-regulation of genes encoding the antioxidant enzymes (*APXa, APXb, CATa, CATb, CATc, SOD2,* and *SOD3*) in PEG primed seeds under nano-ZnO stress. They have concluded that priming with PEG significantly alleviates the toxic effects of nano-ZnO through improved cell structures of leaf and roots.

Seed aging during storage is associated with ROS production. Appearance of oxidative stress results in a decrease of seed quality. Kibinza et al. [161] showed that priming plays an important role in seed recovery from aging through CAT activation. Their results revealed accumulation of hydrogen peroxide (H_2O_2) and reduction of CAT at the gene expression level and protein content during sunflower (*Helianthus annuus* L) seed aging. Interestingly, the adverse results of aging were recovered by seed osmopriming, which led to induction of CAT synthesis by activating gene expression and translation of the enzyme.

Summing up, the management of oxidative status in primed seeds plays a very important role as a machinery, which leads to protection against oxidative stress, recovery from aging, and regulation of ROS production/accumulation. Alleviation in ROS level exerts a signal, which could be perceived, transduced, and crosstalk with other signaling pathways, thus executing physiological response by activation or repression of molecular processes.

9. Reserve mobilization

Recent transcriptome and proteome study of cabbage (*Brassica oleracea*) and *Arabidopsis* seeds as well as integrated transcriptomic and proteomic approach in study of rapeseeds (*Brassica napus*) osmopriming revealed that germination and priming altered similar processes [36, 136, 138]. Indeed, it is proposed that germination-related processes such as respiration, energy metabolism, and early reserve mobilization can also occur during priming [16] (**Figure 2**). Faster and uniform rice (*Oryza sativa*) seed germination due to priming was related to improved activity of α-amylase, resulting in increased level of soluble sugars in primed kernels [168]. Sung and Chang [169] have shown that priming of maize (*Zea mays*) seeds leads to increased activity of enzymes for carbohydrates (α and β amylases) and lipids (isocitrate lyase, ICL) mobilization. Priming of chickpea (*Cicer arietinum* L. Cv PBG-1) seeds with mannitol led to increased activity of amylase, invertases (acid and alkaline), sucrose synthase (SS), and sucrose phosphate synthase (SPS) in shoots of primed seedlings. The higher amylase activity in shoots suggests a rapid hydrolysis of transitory starch formed in the shoots of primed seedlings leading to more availability of glucose for seedling growth. [170]. Higher content of soluble protein, aldolase, and ICL activity has been observed in haloprimed pepper (*Capsicum annum* L) seeds than in control seeds [171]. Moreover, the α-glucosidase accumulation and increased level of globulin degradation products were observed during the priming process of sugar beet (*Beta vulgaris* L.) seed [172]. Similar observation on primed sugar beet seed has been done by Capron et al. [173] who showed increased solubilization of 11S-globulin-β-subunit in response to hydro- and osmopriming. Gallardo et al. [136] have also observed higher polypeptides content in both hydro- and osmoprimed *Arabidopsis* seeds. They were identified as products of 12S-cruciferin-β-subunit degradation.

Kubala et al. [36] have shown that during *Brassica napus* seeds osmopriming and post-priming germination accumulation of transcript and proteins for seed storage proteins occurred. The authors have observed up-accumulation of cruciferin CRU1. The group of six genes encoding GDSL-like lipases, playing a role in triacylglycerols (major storage lipids in rape seeds) catabolism, were strongly up-regulated during post-priming germination while the other three genes for GDSL-like lipases as well as extracellular lipase 6 were up-regulated in osmoprimed seeds [36]. The activation of lipid catabolism-related genes was correlated with the activation of genes involved in lipid transport such as genes encoding bifunctional inhibitor/lipid-transfer protein/seed storage 2S albumin superfamily protein [36]. Priming can also reduce a level of oleosins-proteins, which surround oil bodies. In osmoprimed *Brassica napus* seeds as well as during post-priming germination, down accumulation of oleosin S4-3 protein and down-regulation of *OLEOSIN2* gene were observed, respectively [36].

Activation of respiration and rapid ATP production is primary metabolic events occurring during priming [18], and higher respiratory activity is required to cover energy pool for speed up germination. The increased ATP level/energy charge after priming was observed in tomato (*Solanum lycopersicum*), eggplant (*Solanum melongena*), araucaria (*Araucaria columnaris*), spinach, oat (*Avena sativa*), and cabbage (*Brassica oleracea*) [174, 175]. Primed seed needs a large amount of fuel, which supplies energy required for higher reserve mobilization rate. All together lead to improved energy turnover and increased metabolism rate of primed seed and contribute to better germination and stress tolerance.

10. Holistic "omic" approaches of seed priming

The identification of biomarkers of seed priming effect is a relevant goal for plant physiologists. The sequence of events associated with water-based priming involves a limited hydration step during the soaking period followed by a more or less rapid dehydration phase (**Figure 1**). It may be considered that each phase has its own profile of activation/deactivation. Transcription is not necessarily coordinately associated with translation, resulting in some cases in a limited correspondence between mRNA levels and protein abundance. Some proteins synthesized during incubation may also be degraded during dehydration, although the rate of degradation might somewhat be influenced by the rate of drying. Finally, protein synthesis may also occur as a result of the translation of long-lived mRNA previously synthesized during seed maturation [17, 37, 103, 110].

The advantage of these complementary holistic approaches is that they provide a large set of data allowing physiologist to obtain a global view of all parameters of the metabolism associated with priming. The major disadvantage, however, is that these techniques are rather expensive. As a consequence, samples are frequently analyzed after a given duration of treatment and the time-dependent evolution of parameters are rarely considered. It is still difficult to reconstitute a kinetic approach, but published studies rather provide a large set of information at a given moment while the priming procedure is a dynamic process.

10.1. Transcriptomics

Transcriptomics is the study of the transcriptome—the complete set of RNA transcripts that are produced by the genome under specific circumstances—using high-throughput methods. Several techniques are available for transcriptomic approaches, including microarray, c-DNA amplified fragment length polymorphism (cDNA-AFLP), expressed sequence tag sequencing, serial analysis of gene expression, RNAseq, and massive parallel signature sequencing [176]. According to Buitink et al. [109], more than 1300 genes may be differentially regulated during priming with PEG at -1.7 MPa in *Medicago truncatula*. Genes whose expression are regulated during the priming process are commonly categorized according to the function of the corresponding protein (metabolism and regulation of metabolism, cell cycle regulation, DNA processing, transcription regulation, cellular transport and communications, stress responses, etc.). However, a consistent part of the identified genes is still not annotated.

In *Brassica napus*, numerous priming-regulated genes are involved in gluconeogenesis, which is essential for triacylglycerol breakdown into small molecules. Another important category of genes involved in water-based priming of rapeseed encodes for transcription factors, and this is especially the case after hormopriming with ABA [138]. In the same species, it was demonstrated that germination of primed seeds involves a specific set of genes comparatively to germination of unprimed ones [36]. In young seedlings obtained from primed seeds, the expression of stress-related genes is often more rapid than in plants from unprimed ones, as recently exemplified for cold tolerance [26]. Not only genes related to protein synthesis but also genes involved in protein degradation may be induced by priming. Some transcription factors may be down-regulated during soaking and up-regulated during drying [36]. Genes involved in the synthesis of osmoprotectant, such as proline, may also be regulated at various steps of priming [120].

In *Brassica oleracea*, transcripts that are abundant in dry seeds rapidly decline during osmo-priming and germination expression programs are initiated during osmopriming [138]. For these authors, genes expressed during slow drying following the soaking period are correlated with the stress resistance properties of primed-material. This may especially concern cell-cycle-regulated genes, enzymes involved in C metabolism, and components of the translation machinery. Acquisition of the desiccation tolerance after the soaking period is a crucial trait for priming. Transcriptomic data demonstrated that the molecular determinism associated with such trait may be, at least partly, similar to those involved during a « normal » seed dehydration process during maturation. In *Medicago truncatula*, several regulatory genes expressed during drought stress are also up-regulated during seed maturation. During priming, these genes are involved in post-soaking dehydration while genes conditioning cell cycle, cell wall biogenesis, and energy metabolism are repressed [109]. Transcripts accumulated after PEG-priming treatment comprise those encoding for LEA and seed storage proteins, or genes controlling seed dormancy while genes involved in photosynthesis and cell wall modifications are commonly repressed [108].

Beside involvement of transcription factors, it could be of primary importance in the future to analyze the priming-induced epigenetic changes. Indeed, DNA methylation may be directly involved in « stress memory » and some changes, such as cytosine methylation or overexpression of histone deacetylase was reported to occur during germination [78].

10.2. Proteomics

Proteome is the entire set of proteins present at a given moment in a given biological sample and proteomic is the large-scale study allowing identification and quantification of these proteins. Proteomics has now been practiced in different plant systems through the separation of proteins by two-dimensional gel electrophoresis followed by peptide mass fingerprinting. Tandem mass spectrometry can get sequence information from individual isolated peptides. Quantitative proteomics represent an important extension enabling the comparison of changes in protein levels across different samples [177]. Gel-free shotgun proteomic is an alternative for identification and quantification of protein in large-scale studies. This method was recently used to reveal potential biomarkers of priming-induced salt tolerance in durum wheat with a concomitant use of protein fractionation and hydrogel nanoparticles enrichment technique [27]. According to these authors, hydropriming was accompanied by a significant change in 72 proteins. Most of them are involved in proteolysis, protein synthesis, metabolism regulation, and disease or defense response. Priming with ascorbic acid changed the pattern of proteome signature and most of identified proteins are involved in metabolism regulation, antioxidant protection, repair processes, and methionine-related metabolism.

Cycloheximide, a potent inhibitor of protein synthesis, may be added to priming or to germination solution to analyze the precise requirement of protein synthesis during these steps. It appears that protein synthesis is not necessarily required for normal germination process, which could be related to the fact that in some cases, germination relies only on translation of pre-stored long-lived mRNA produced during seed maturation [14, 178]. Cycloheximide may drastically affect osmopriming-induced improvement of Cd resistance in *Poa pratensis* and *Trifolium repens* in relation to a decrease in α and β-amylase [179]. Protein concentration in primed seeds is the result of two contrasting pathways: the first implies specific protein synthesis while the second involves the breakdown of storage proteins through protease activation. It was demonstrated that priming may act on protease activities and also influence the expression of protease encoding-gene [180].

Some strategies are also used to separately identify proteins in embryos and in embryo-surrounding tissues. In monocots, a special attention is often paid to aleurone layers involved in hydrolytic enzyme synthesis [181]. Aleurone layer is easily separated in some plant material such as wheat or barley but is more difficult with other monocot such as turfgrasses with very small seeds. In wheat, it was demonstrated that the set of proteins regulated by priming is quite different in embryos and in surrounding tissues and the number of proteins affected by priming was by far higher in the embryos than in the other parts of the seed. Proteins regulated in embryos involve those directly linked to metabolism regulation (especially methionine biosynthesis, glutamate/glutamine metabolism, amino acids synthesis, etc.), energy supply, cell growth and maintenance of cell structure while proteins up-regulated in surrounding tissues are mainly involved in reserves mobilization or in the management of the oxidative stress [67]. In barley, however, ascorbate peroxidase was observed only in the embryo while several other redox-related proteins differed in spatio-temporal patterns at the onset of radicle elongation [181].

Priming is considered as an invigoration treatment and several proteomic approaches were performed to unravel the biomarker of seed vigor. Catusse et al. [182] found that 18 proteins are accumulated during hydropriming of sugarbeet seeds and that the same proteins appear directly reduced during aging. Seed vigor appears directly related to lipid and starch mobilization, protein synthesis and methionine cycle. In poplar, more than 81 proteins showed a significant change in abundance when comparing the proteomes among seed with different vigor [183]. According to these data, the decrease in seed vigor is an energy-dependent process, which requires protein synthesis and degradation as well as cellular defense and rescue. Salicylic acid (SA) was proposed as an invigorating elicitor promoting seed germination under saline conditions: SA re-induced the late maturation program during early stages of germination, induced the synthesis of antioxidant enzymes, and improved the quality of protein translation [184]. Similarly, the "prime-ome" approach performed by Tanou et al. [185] confirmed the importance of redox proteomic and processes such as N-nitrosilation, tyrosine nitration, and mitogen-activated protein kinase MPK3 signaling in priming effects.

Osmotic priming was reported to trigger desiccation tolerance in *Medicago truncatula* [110]. Proteomic analysis demonstrated that such trait is directly linked to the synthesis of late-embryogenesis abundant proteins from different groups. Secondary structure of some proteins was compared in the hydrated and dry state after fast or slow drying using Fourier transform infrared spectroscopy, which confirms that these proteins adopted α helices and β-sheets conformation during drying process.

Some proteomic approaches are conducted on plant species whose genomes are not sequenced. A recent 2DE-MS/MS-based proteomic study was conducted on pearl millet seeds primed by β-aminobutyric acid and showed an over-representation of proteins belonging to glucose metabolism, and a majority of induced proteins are directly related to energy [106]. Seedlings issued from those seeds are more resistant to downy mildew (*Sclerospora graminicola*). It is interesting to note that several of the elicited proteins are present in the extracellular space and in organelles (mainly mitochondrion and chloroplast)

10.3. Metabolomics

Metabolome refers to the complete set of small-molecule metabolites present within a plant tissue or organ at a moment. Metabolomic should thus be considered as the quantitative measurements of the whole set of compounds involved in the metabolism of a given biological sample. Plant metabolomics has become an essential part of functional genomics. It often appears difficult to analyze the entire range of metabolites by a single analytical method, and several tools are thus commonly combined for this purpose: high performance liquid chromatography (HPLC), gas chromatography (GC), electrospray ionization coupled with mass spectrometry, capillary electrophoresis, atmospheric pressure chemical ionization (APCI), and secondary ion mass spectrometry (SIMS). Metabolomic approach, however, still remains difficult to perform on seed material considering the high proportion of molecule issued from reserve mobilization. An excess of soluble sugar in graminaea and an excess of small lipids issued from oil digestion in oleaginous plant material may greatly hamper isolation of other compounds involved in hastening the germination of primed seeds. Similarly, some important compounds might be present only in specific tissues or cell compartments and react with sugars during extraction processes; this is especially the case of gibberellins which play a crucial role in germination but are believed to conjugate with sugar or phenol compounds, which greatly compromise isolation of and identification procedures of numerous metabolites [186].

Priming process may drastically modify the synthesis and accumulation of endogenous antioxidant such as glutathione, ascorbic acid, and even α-tocopherol. A modification in reducing sugars concentration, such as glucose resulting from partial starch digestion during priming, may influence protein glycation, a nonenzymatic reaction between reducing sugars and amino groups in protein [150]. The amount of proline was reported to be modified in osmoprimed seeds [120]. Some other data reported modifications in aspartate, leucine, threonate, glutamate, fumarate, or pinitol content in primed seeds from different species [1, 2, 35, 104, 152]. Methionine may also play a key role in priming process, mainly as a precursor of both ethylene and polyamine synthesis. Polyamines are small aliphatic molecules influencing all aspects of plant metabolism and development, including the germination processes. Ethylene is also involved in various aspects of seed germination, and priming was reported to modify the kinetics of ethylene synthesis from its precursor 1-aminocyclopropane-1-carboxylic acid (ACC) [186]. Other phytohormones are expected to assume important functions in priming effect (ABA, gibberellins, cytokinins, auxin, etc.) [50, 52], but the quantitative impact of priming procedure on these compounds is still not well established.

11. Main limitations of priming techniques

Seed priming has emerged as an effective approach for increasing seed vigor. The optimal treatment differs between species, cultivar, and seed lots. Such variability is a major limitation of the priming method since numerous trials are required to identify the most appropriate strategy for each situation. There is no "general rule" concerning modalities of seed priming and there is no clear trend of priming response according to the taxonomic position of the species [188]. This, undoubtedly, limits the commercial implementation of priming treatments.

Some priming treatments may imply a risk of medium contamination by fungi and bacteria, which may heavily impair subsequent seed germination [189]. This may especially be the case for PEG and sometimes requires the simultaneous use of pesticide, although the impact of these compounds on the priming efficiency remains unknown. After priming treatment, seeds are dried back to their initial moisture content, but this dehydration phase is usually performed rapidly and occurs faster than classical dehydration of maturing seeds. It has been hypothesized that this brutal desiccation procedure alters the beneficial effect of priming [2].

The major drawback of priming is that it reduces the longevity of primed seeds as compared with the nonprimed seeds [190–192]. Storability of primed seed material is consequently reduced, and this results in higher costs for material maintenance for farmers and seed companies. The loss of viability, however, appears quite variable depending on the species, cultivars, and seed lots. In extreme cases, priming-induced advantages may even disappear after only 14 d of storage and the obtained seedling may then perform worse than those issued from unprimed seeds [193]. One of the limitations of these studies, however, is that experiments are commonly performed on artificially aged seeds using short-term exposure to high temperature in a moist environment, which is not necessarily fulling relevant from a real aging process. Some studies, using classical long-term strategies, reported that longevity is not necessarily affected or may even be increased by priming treatments [194]. Hussain et al. [195] recently demonstrated that post-priming temperature plays a key role in maintenance of seed longevity, which is indeed rapidly compromised in rice seeds maintained at room temperature while it remains intact when the material is stored at 4°C. According to these authors, the deleterious effect observed at 25°C storage was related to hampered starch metabolism in primed seeds. It was also suggested that maintenance of the seed longevity at 4°C may be due to a high viscosity that strongly reduces molecular mobility in the cytoplasm and thus limits the impact of deteriorative process even in seeds exhibiting low water content [196]. Oxygen during storage may trigger metabolic processes in primed seeds, which did not re-establish a true quiescent stage after dehydration when stored at room temperature while oxygen had no influence at low temperatures [195].

In some cases, repeat priming treatment after storage may partly remove the damaging effect on seed viability [193] while, in other cases, such a loss is permanent and not reversible [194]. Whatever, the fact that an additional treatment may be required to restore full germination potential represents an additional cost and source of variability.

12. Conclusions

Seed priming is an old empirical strategy used since centuries by farmers, and since decades by seed companies, to improve germination processes in cultivated plant species. The underlying mechanisms involved in this positive impact of pre-sowing treatments remained obscure for a long time. The present review aimed to summarize recent information provided by various tools allowing the identification of molecular cues conditioning priming efficiency.

Data obtained from molecular approaches applied to some well-known plant species (rice, rapeseed, tomato, etc.) are now available. The role of genes associated to metabolic and cell cycle events now starts to be deciphered, mainly those encoding for translational components such as ribosomal subunits, translational initiation factor, enzyme involved in carbon metabolism, histones, and transcription factors. A putative epigenetic basis of priming effects should also be considered. Some authors identified enzyme activity changes in relation to priming while others also reported numerous changes in the seed storage protein. Priming has also been reported to increase proteins related to the cell cycle activities such as α- and δ-DNA polymerase. Protecting proteins like dehydrins or HSP is expected to assume protective functions during the dehydration step. Similarly, proteins involved in water transport, cell wall modification, cytoskeletal organization, and cell divisions, may be to some extent regulated during priming. Gene expression and enzyme activities involved in osmocompatible solute synthesis may be of primary importance to regulate tissue protection during the dehydration step and water fluxes during the final germination phase. Measurements of water uptake by primed seeds suggest a reduction in the lag time of imbibition. Water uptake and its subsequent cell-to-cell movement during germination might be controlled by aquaporins and expression of the corresponding genes constitutes a specific target of the priming treatment.

Among the different hypothesis proposed to explain the biochemical basis for germination improvement, DNA replication and cell cycle advancement during priming treatment as well as synchronization of the cell cycle at the G2 phase are supported by some experimental evidences. DNA synthesis is involved during priming treatment itself but also during post-germinative events. Changes are also observed with the modification of the membrane structure and reorganization of mitochondrial integrity. Activation of antioxidative properties by priming treatment may also explain the improved behavior of plant material, especially when final germination and/or growth occur under stress condition.

The priming-induced decrease of the storage capacity is a major limitation for the application of the priming technique by seed companies. Partial vacuum storage may be useful for extending the longevity of primed seeds. Improved longevity may be related to enhanced antioxidative activity that minimizes the accumulation of total peroxide during long-term storage. Another challenge for private seeds company is to identify appropriate treatments able to restore vigor of old dry seed lots in order to increase their mean percentage of germination to values compatible with commercial purposes.

If priming may undoubtedly be considered as a valuable strategy to improve stand establishment, its impact on final yield and crop production has not be always confirmed. Most studies devoted to "omics" approaches of seed priming are performed on young seedling cultivated under fully controlled environmental conditions. The link between seedling behavior in plant growth chambers and the adult plant performance in field conditions is far from being clear. As a consequence, there is an urgent need to focus on transcriptomic, proteomic, and metabolomics of adult plants issued from primed seeds, especially at the reproductive stage, in order to assess the long-term impact of priming on cultivated plants throughout the plant cycle.

In cultivated plant species, a given priming treatment also has contrasting effects on various cultivars. It may be hypothesized that the ability to respond to priming treatment might be genetically controlled but, to the best of our knowledge, no data are available concerning this important aspect. Thus, further progresses are needed not only to identify the set of genes that are regulated by priming, but also the set of genes that putatively regulate priming response and efficiency themselves.

Author details

Stanley Lutts[1*], Paolo Benincasa[2], Lukasz Wojtyla[3], Szymon Kubala S[3], Roberta Pace[2], Katzarina Lechowska[3], Muriel Quinet[1] and Malgorzata Garnczarska[3]

*Address all correspondence to: stanley.lutts@uclouvain.be

1 Groupe de Recherche en Physiologie végétale (GRPV), Earth and Life Institute-Agronomy, Université catholique de Louvain, Louvain-la-Neuve, Belgium

2 Department of Agricultural and Environmental Sciences, University of Perugia, Perugia, Italy

3 Faculty of Biology, Department of Plant Physiology, Adam Mickiewicz University, Poznan, Poland

References

[1] Chen K, Arora R. Dynamics of the antioxidant system during seed osmopriming, post-priming germination, and seedling establishment in spinach (*Spinacia oleracea*). Plant Science. 2011;180:212-220. DOI:10.1016/j.plantsci.2010.08.007

[2] Paparella S, Araújo SS, Rossi G, Wijayasinghe M, Carbonera D, Balestrazzi A. Seed priming: state of the art and new prospectives. Plant Cell Reports. 2015;34:1281-1293. DOI: 10.1007/s00299-015-1784-y

[3] Zhang F, Yu J, Johnston CR, Wang Y, Zhu K, Lu F, Zhang Z, Zou J. Seed priming with polyethylene glycol induces physiological changes in sorghum (*Sorghum bicolor* L. Moench) seedlings under suboptimal soil moisture environments. PLoS One. 2015;10: e0140620. DOI: 10.1371/journal.pone.0140620

[4] Santini BA, Martorell C. Does retained-seed priming drive the evolution of serotiny in drylands? An assessment using the cactus *Mammilaria hernandezii*. American Journal of Botany. 2013;100:365-373.

[5] Balmer A, Pastor, Gamir J, Flors V, Mauch-Mani B. The « prime-ome »: towards a holistic approach to priming. Trends in Plant Science. 2015;20:443-452.

[6] Tanou G, Fotopoulos V, Molassiotis A. Priming against environmental challenges and proteomics in plants: update and agricultural perspectives. Frontiers in Plant Science. 2012;3:216.

[7] Gamir J, Sánchez-Bel P, Flors V. Molecular and physiological stages of priming: how plants prepare for environmental challenges. Plant Cell Reports. 2014;33:1935-1949.

[8] Evenari M. Seed physiology: its history from Antiquity to the beginning of the 20th century. Botanical Review. 1984;50:119-142

[9] Evelyn JS. A Discourse of Forest Trees and the Propagation of Timber. John Martin (Royal Society). London; 1664.

[10] Ingenhousz J. Experiments Upon Vegetables. 1779. P. Elmsly and H. Payne, London.

[11] Amici GB. Note on pollen action mode on stigmata. Annales Sciences Naturelles Botanique. 1830;21:329-332

[12] Sachs J. About treatments modifying germination strength in seeds. Botanical Zeitung. 1859;17:177-188.

[13] Sachs J. Vorlesungen Lectures on Plant Physiology. Berlin; 1887.

[14] Khan AA, Peck NH, Samimy C. Seed osmoconditioning: physiological and biochemical changes. Israel Journal of Botany. 1981;29:133-144.

[15] Hussain M, Farooq M, Basra SMA, Ahmad N. Influence of seed priming techniques on the seedling establishment, yield and quality of hybrid sunflower. International Journal of Agriculture & Biology. 2006;8:14-18. DOI: 1560-8530/2006/08-1-14-18

[16] Varier A, Vari AK, Dadlani M. The subcellular basis of seed priming. Current Science. 2010;99: 450-456.

[17] Taylor AG, Allen PS, Bennett MA, Bradford JK, Burris JS, Mishra MK. Seed enhancements. Seed Science Research. 1998;8:245-256. DOI: 10.1017/S0960258500004141

[18] McDonald MB. Seed priming. In: Black M, Bewley JD, editors. Seed Technology and its Biological Basis. Sheffield, Sheffield Academic Press; 2000. p 287-325.

[19] Rahman MM, Ahammad KU, Alam MM. Effect of soaking condition and temperature on imbibition rate of maize and chickpea seeds. Research Journal of Seed Science. 2011;4:117-124. DOI: 10.3923/rjss.2011.117.124

[20] Basra SMA, Zia MN, Mahmood T, Afzal I, Khaliq A. Comparison of different invigoration techniques in wheat *(Triticum aestivum* L.) seeds. Pakistan Journal of Arid Agriculture. 2002;5:11-16.

[21] Srivastava AK, Lokhande VH, Patade VYwx, Suprasanna P, Sjahril R, D'Souza SF. Comparative evaluation of hydro-, chemo-, and hormonal priming methods for imparting salt and PEG stress tolerance in Indian mustard (*Brassica juncea* L.). Acta Physiologiae Plantarum. 2010;32:1135-44. 10.1007/s11738-010-0505-y

[22] Omidi H, Khazaei F, Hamzi Alvanagh S, Heidari-Sharifabad H. Improvement of seed germination traits in canola (*Brassica napus* L.) as affected by saline and drought stresses. Plant Ecophysiology. 2009;3:151-158.

[23] Kaya MD, Okçu G, Atak M, Çıkılı Y, Kolsarıcı Ö. Seed treatments to overcome salt and drought stress during germination in sunflower (*Helianthus annuus* L.). European Journal of Agronomy. 2006;24: 291-295. Doi:10.1016/j.eja.2005.08.001

[24] Goswami A, Banerjee R, Raha S. Drought resistance in rice seedlings conferred by seed priming: role of the anti-oxidant defense mechanisms. Protoplasma. 2013;250:1115-29. DOI: 10.1007/s00709-013-0487-x

[25] Posmyk MM, Janas KM. Effects of seed hydropriming in presence of exogenous proline on chilling injury limitation in *Vigna radiata* L. seedlings. Acta Physiologiae Plantarum. 2007;29:509-517. DOI: 10.1007/s11738-007-0061-2

[26] Patade VY, Khatri D, Manoj K, Kumari M, Ahmed Z. Cold tolerance in thiourea primed capsicum seedlings is associated with transcript regulation of stress responsive genes. Molecular Biology Reports. 2012;39:10603-10613.

[27] Fercha A, Capriotti AL, Caruso G, Cavaliere C, Gherroucha H, Samperi R, Stampachiacchierea S, Laganaa A. Gel-free proteomics reveal potential biomarkers of priming-induced salt tolerance in durum wheat. Journal of Proteomics. 2013;91:486-499. DOI: 10.1016/j.jprot.2013.08.010

[28] Rowse HR. Methods of priming seeds. 1992. United States Patent No. 5,119,589.

[29] Rowse HR. Methods of priming seeds. 1991. UK Patent No. 2192781.

[30] Warren JE, Bennett MA. Seed hydration using the drum priming system. HortScience. 1997;32:1220-1221.

[31] Da Silva CB, Marcos-Filho J, Jourdan P, Bennett MA. Performance of bell pepper seeds in response to drum priming with addition of 24-Epibrassinolide. HortScience. 2015;50:873-878.

[32] Harris D, Joshi A, Khan PA, Gothkar P, Sodhi, PS. On-farm seed priming in semi-arid agriculture: development and evaluation in maize, rice and chickpea in India using participatory methods. Experimental Agriculture. 1999;35:15-29.

[33] Harris D, Raghuwanshi BS, Gangwar JS, Singh SC, Joshi KD, Rashid I, Hollington PI. Participatory evaluation by farmers for on-farm seed priming in wheat in India, Nepal and Pakistan. Experimental Agriculture. 2001;37:403-415.

[34] Harris D, Breese WA, Kumar Rao JVDK. The improvement of crop yield in marginal environments using 'on-farm' seed priming: nodulation, nitrogen fixation and disease resistance. Australian Journal of Agricultural Research. 2005;56:1211-1218. DOI: 10.1071/AR05079

[35] Di Girolamo G, Barbanti L. Treatment conditions and biochemical processes influencing seed priming effectiveness. Italian Journal of Agronomy. 2012;7:8-18. DOI: 10.4081/ija.2012.e25

[36] Kubala S, Garnczarska M, Wojtyla Ł, Clippe A, Kosmala A, Żmieńko A, Lutts S, Quinet M. Deciphering priming-induced improvement of rapeseed (*Brassica napus* L.) germination through an integrated transcriptomic and proteomic approach. Plant Science. 2015;231:94-113. DOI: 10.1016/j.plantsci.2014.11.008

[37] Yacoubi R, Job C, Belghazi M, Chaibi W, Job D. Proteomic analysis of the enhancement of seed vigour in osmoprimed alfalfa seeds germinated under salinity stress. Seed Science Research. 2013;23:99-110. DOI: 10.1017/S0960258513000093

[38] Salah SM, Yajing G, Dongdong C, Jie L, Aamir N, Qijuan H, Weimin H, Mingyu N, Jin H. Seed priming with polyethylene glycol regulating the physiological and molecular mechanism in rice (*Oryza sativa* L.) under nano-ZnO stress. Scientific Reports. 2015;5:14278. DOI: 10.1038/srep14278

[39] Mereddy R, Wu L, Hallgren SW, Conway KE. Solid matrix priming improves vigor of okra seeds. Proceedings of the Oklahoma Academy of Science. 2000;80:1-5.

[40] Singh PK., Pandita VK, Tomar BS, Seth R. Standardization of priming treatments for enhancement of seed germination and field emergence in carrot. Indian Journal of Horticulture. 2015;72:306- 309. DOI: 10.5958/0974-0112.2015.00059.6

[41] Mercado MFO, Fernandez PG. Solid matrix priming of soybean seeds. Philippine Journal of Crop Science. 2002;2:27-35.

[42] Kępczyńska E, Piękna-Grochała J, Kępczyński J. Effects of matriconditioning on onion seed germination, seedlings emergence and associated physical and metabolic events. Plant Growth Regulation. 2003;41:269-278.

[43] Zhang CF, Hu J, Lou J, Zhang Y, Hu WM. Sand priming in relation to physiological changes in seed germination and seedling growth of waxy maize under high-salt stress. Seed Science and Technology. 2007;35:733-738.

[44] Andreoli C, Andrade RV. Integrating matriconditioning with chemical and biological seed treatments to improve vegetable crop stand establishment and yield under tropical conditions. Seed Technology. 2002;24:89-99.

[45] Ilyas S, Sutariati GA, Suwarno FC, Sudarsono S. Matriconditioning improved quality and protein level of medium vigor hot pepper seed. Seed Technology. 2002;24:67-77.

[46] Pandita VK, Anand A, Nagarajan S, Seth R, Sinha SN. Solid matrix priming improves seed emergence and crop performance in okra. Seed Science and Technology. 2010;38:665-674. DOI: 10.15258/sst.2010.38.3.14

[47] Galhaut L, Lespinay A, Walker DJ, Bernal MP, Correal E, Lutts S. Seed priming of *Trifolium repens* L. improved germination and early seedling growth on heavy metal-

contaminated soil. Water Air Soil Pollution. 2014;225:1-15. DOI: 10.1007/s11270-014-1905-1

[48] Iqbal M, Ashraf, M. Gibberellic acid mediated induction of salt tolerance in wheat plants: growth, ionic partitioning, photosynthesis, yield and hormonal homeostasis. Environmental and Experimental Botany. 2013;86:76-85. DOI: 10.1016/j.envexpbot.2010.06.002

[49] Iqbal M, Ashraf M, Jamil A. Seed enhancement with cytokinins: changes in growth and grain yield in salt stressed wheat plants. Plant Growth Regulation. 2006;50:29-39. DOI: 10.1007/s10725-006-9123-5

[50] Zheng M, Tao Y, Hussain S, Jiang Q, Peng S, Huang J, Cui K, Nie L. Seed priming in dry direct-seeded rice: consequences for emergence, seedling growth and associated metabolic events under drought stress. Plant Growth Regulation. 2016;78:167-178. DOI: 0.1007/s10725-015-0083-5

[51] Farooq M, Wahid A, Lee DJ. Exogenously applied polyamines increase drought tolerance of rice by improving leaf water status, photosynthesis and membrane properties. Acta Physiologia Plantarum. 2009;31:937-945. DOI: 10.1007/s11738-009-0307-2

[52] Afzal I, Basara SMA, Farooq M, Nawaz A. Alleviation of salinity stress in spring wheat by hormonal priming with ABA, salicylic acid and ascorbic acid. International Journal of Agriculture and Biology. 2006;8:23-28.

[53] Callan NW, Marthre DE, Miller JB. Bio-priming seed treatment for biological control of *Pythium ultimum* preemergence damping-off in sh-2 sweet corn. Plant Disease. 1990;74: 368-372.

[54] Reddy PP. Bio-priming of seeds. In: Reddy PP, editor. Recent Advances in Crop Protection. India, Springer; 2013. p 83-90. DOI: 10.1007/978-81-322-0723-8_6

[55] Callan NW, Mathre DE, Miller J.B, Vavrina CS. Biological seed treatments: factors involved in efficacy. Horticultural Science.1997;32:179-183.

[56] Müller H, Berg G. Impact of formulation procedures on the effect of the biocontrol agent *Serratia plymuthica* HRO-C48 on Verticillium wilt in oilseed rape. BioControl. 2008;53: 305-316. DOI 10.1007/s10526-007-9111-3

[57] Gulick BR. Plant Growth-Promoting Bacteria: Mechanisms and Applications. Hindawi Publishing Corporation, Scientifica. 2012;1-15. DOI: 10.6064/2012/963401

[58] Timmusk S, Abd El-Daim IA, Copolovici L, Tanilas T, Kännaste A, Behers L, Nevo E, Seisenbaeva G, Stenström E, Niinemets Ü. Drought-tolerance of wheat improved by rhizosphere bacteria from harsh environments: enhanced biomass production and reduced emissions of stress volatiles. PLoS One. 2014;9(5):e96086. DOI: 10.1371/journal.pone.0096086

[59] Raj NS, Shetty NP, Shetty HS. Seed bio-priming with *Pseudomonas fluorescens* isolates enhances growth of pearl millet plants and induces resistance against downy mildew. International Journal of Pest Management. 2004;50:41-48. DOI: 10.1080/09670870310001626365

[60] Kaymak HC, Guvenc I, Yarali F, Donmez MF. The effects of bio-priming with PGPR on germination of radish (*Raphanus sativus* L.) seeds under saline conditions. Turkish Journal of Agriculture and Forestry. 2009;33:173-179. DOI: 10.3906/tar-0806-30

[61] Khaliq A, Aslam F, Matloob A, Hussain S, Geng M, Wahid A, Rehman H. Seed priming with selenium: consequences for emergence, seedling growth, and biochemical attributes of rice. Biological Trace Element Research. 2015;166:236-244. DOI:10.1007/s12011-015-0260-4

[62] Guan YJ, Hu J, Wang XJ, Shao CX. Seed priming with chitosan improves maize germination and seedling growth in relation to physiological changes under low temperature stress. Journal of Zhejiang University Science B. 2009;10:427-443. DOI: 10.1631/jzus.B0820373

[63] Duan P, Ding F, Wang F, Wang BS. Priming of seeds with nitric oxide donor sodium nitroprusside (SNP) alleviates the inhibition on wheat seed germination by salt stress. Journal of Plant Physiology and Molecular Biology. 2007;33:244-250.

[64] Anosheh HP, Sadeghi H, Emam Y. Chemical priming with urea and KNO_3 enhances maize hybrids (*Zea mays* L.) seed viability under abiotic stress. Journal of Crop Science and Biotechnology. 2011;14:289-295. DOI: 10.1007/s12892-011-0039-x

[65] Jisha KC, Puthur JT. Seed priming with BABA (β-amino butyric acid): a cost-effective method of abiotic stress tolerance in *Vigna radiata* (L.) Wilczek. Protoplasma. 2015. DOI: 10.1007/s00709-015-0804-7

[66] Farooq M. Irfan M, Aziz Y, Ahmad I, Cheema SA. Seed priming with ascorbic acid improves drought resistance of wheat. Journal of Agronomy and Crop Science. 2013;199:12-22. DOI: 10.1111/j.1439-037X.2012.00521.x

[67] Fercha A, Capriotti AL, Caruso G, Cavaliere C, Samperi R, Stampachiacchiere S, Laganà A. Comparative analysis of metabolic proteome variation in ascorbate-primed and unprimed wheat seeds during germination under salt stress. Journal of Proteomics. 2014;108:238-25. DOI:10.1016/j.jprot.2014.04.040

[68] Farooq M, Wahid A, Siddique KHM. Micronutrients application through seed treatments – a review. Journal of Soil Science and Plant Nutrition. 2012;12:125-142. DOI: 10.4067/S0718-95162012000100011

[69] Arif M, Waqas M, Nawab K, Shahid M. Effect of seed priming in Zn solutions on chickpea and wheat. African Crop Science Conference Proceedings. 2007;8:237-240.

[70] Abbas SQ, Hassan MU, Hussain B, Rasool T, Ali Q. Optimization of zinc seed priming treatments for improving the germination and early seedling growth of *Oryza sativa*. Advancements in Life Sciences. 2014;2:31-37.

[71] Imran M, Mahmood A, Romheld V, Neumann G. Nutrient seed priming improves seedling development of maize exposed to low root zone temperatures during early growth. European Journal of Agronomy. 2013;49:141-148. DOI: 10.1016/j.eja.2013.04.001

[72] Shaheen HL, Iqbala M, Azeema M, Shahbazb M, Shehzadia M. K-priming positively modulates growth and nutrient status of salt-stressed cotton (*Gossypium hirsutum*) seedlings. Archives of Agronomy and Soil Science. 2015. DOI: 10.1080/03650340.2015.1095292

[73] Jalali AH, Selehi F. Sugar beet yield as affected by seed priming and weed control. Archives of Agronomy and Soil Science. 2013;59:281-288.

[74] Rashid A, Harris D, Hollington P, Ali S. On-farm seed priming reduces yield losses of mungbean (*Vigna radiata*) associated with mungbean yellow mosaic virus in the North West Frontier Province of Pakistan. Crop Protection. 2004;23:1119-1124.

[75] El-Araby MM, Moustafa SMA, Ismail AI, Hegazi AZA. Hormones and phenol levels during germination and osmopriming of tomato seeds, and associated variation in protein patterns and anatomical seed features. Acta Agronomica Hungarica. 2006;54:441-458.

[76] SchwemberAR, Bradford KJ. A genetic locus and gene expression pattern associated with the priming effect on lettuce seed germination at elevated temperature. Plant Molecular Biology. 2010;73:105-118.

[77] Sung Y, Cantliffe DJ, Nagata RT, Nascimiento WM. Structural changes in lettuce seed during germination at high temperature altered by genotype, seed maturation temperature and seed priming. Journal of the American Society of Horticultural Science. 2008;133:300-311.

[78] Chen K, Arora R. Priming memory invokes seed stress-tolerance. Environmental and Experimental Botany. 2013;94:33-45.

[79] Sun H, Lin L, Wang X, Wu S, Wang X. Ascorbate-glutathione cycle of mitochondria in osmoprimed soybean cotyledons in response to imbibitional chilling injury. Journal of Plant Physiology. 2011;168:226-232.

[80] Nascimento WM, Huber DJ, Cantliffe DJ. Carrot seed germination and respiration at high temperature in response to seed maturity and priming. Seed Science and Technology. 2013;41:164-169.

[81] Kanto U, Jutamanee K, Osotsapar Y, Chai-Arree W, Jattupornpong S. Promotive effect of priming with 5-aminolevulinic acid on seed germination capacity, seedling growth

and antioxidant enzyme activity in rice subjected to accelerated ageing treatment. Plant Production Science. 2015;18:443-454.

[82] Singh A, Gupta R, Pandey R. Rice seed priming with picomolar rutin enhances rhiospheric *Bacillus subtilis* CIM colonization and plant growth. PLoS One. 2016;11:e0146013.

[83] Hameed A, Farooq T, Hameed A, Ibrahim M, Sheikh MA, Basra SMA. Wheat seed germination, antioxidant enzymes and biochemical enhancements by sodium nitroprusside priming. Agrochimica. 2015;59:93-107.

[84] Pouramir-Dashtmian F, Khajel-Hosseini M, Esfahani M. Improving chilling tolerance of rice seedling by seed priming with salicylic acid. Archives of Agronomy and Soil Science. 2014;60:1291-1302.

[85] Muhammad I, Kolla M, Volker R, Günter N. Impact of nutrient seed priming on germination, seedling development, nutritional status and grain yield of maize. Journal of Plant Nutrition. 2015;38:1803-1821.

[86] Ahmed M, Qadeer U, Ahmed ZI, Hassan FU. Improvement of wheat (*Triticum aestivum*) drought tolerance by seed priming with silicon. Archives of Agronomy and Soil Science. 2016;62:299-315.

[87] Taylor A, Seel WE. Do *Striga hermonthica*-induced changes in soil matric potential cause the reduction in stomatal conductance and growth of infected maize? New Phytologist. 1998;138:67-73.

[88] Zelonka L, Veneranda S, Māra V. Effect and aftereffect of barley seed coating with hoshors on germination, photosynthetic pigments and grain yield. Acta Universitatis Ltviensis Biology. 2005;691:111-119.

[89] Jamil M, Charnikhova T, Verstappen F, Ali Z, Wainwright H, Bouwmeester H. Effect of phosphate-based seed priming in strigolactone production in *Striga hermonthica* infection in cereals. Weed Research. 2014;54:207-313.

[90] Pame AR, Kreye C, Johnson D, Heuer S, Becker M. Effects of genotype, seed P concentration and seed priming on seedling vigor of rice. Experimental Agriculture. 2015;51:370-381.

[91] Atique-ur-Rehman M, Farooq M, Cheema ZA, Wahid A. Seed priming with boron improves growth and yield of fine grain aromatic rice. Plant Growth Regulation. 2012;68:189-201.

[92] Liu Z, Ma Z, Guo X, Shao H, Cui Q, Song W. Changes of cytosolic Ca^{2+} fluorescent intensity and plasma membrane calcium channels of maize root tip cells under osmotic stress. Plant Physiology and Biochemistry. 2010;48:860-865.

[93] Khan AA, Peck NH, Taylor AG, Samimy C. Osmoconditioning of beet seed to improve emergence and yield in cold soils. Agronomy Journal. 1983;75:788-794.

[94] Musa AM, Harris D, Johansen C, Kumar J. Short duration chickpea to replace fallow after aman rice: the role of on-farm seed priming in the High Barind Tract of Bangladesh. Experimental Agriculture. 1981;37:509-521.

[95] Murungu FS, Chiduza C, Nyamugafa P, Clarck LJ, Whalley WR, Finch-Savage WE. Effect of on-farm seed priming on consecutive daily sowing occasions on the emergence and growth of maize in semi-arid Zimbabwe. Field Crop Research. 2004;89:49-57.

[96] Shah Z, Haq IU, Rehman A, Khan A, Afzal M. Soil amendments and seed priming influence nutrients uptake, soil properties, yield and yield-components of wheat (*Triticum aestivum* L.) in alkali soils. Soil Science and Plant Nutrition. 2013;59:262-270.

[97] Binang WB, Shiyam JQ, Ntia JD. Effect of seed priming method on agronomic performances and cost effectiveness of rainfed, dry-seeded NERICA rice. Research Journal of Seed Science. 2012;5:136-143.

[98] Priya P, Patil VC, Aravind Kumar BN. Effect of seed priming practices on growth, yield and economics of maize (*Zea mays* L.) based cropping systems under rainfed conditions in Northern Karnataka, India. Plant Archives 2013;1:207-213.

[99] Patil BD, Manjare MR. Effect of seed priming on germination and bulb yield in onion (*Allium cepa* L.). Ecology, Environment and Conservation. 2013;19:23-246.

[100] Sharma AD, Rathore SVS, Srinivasan K, Tyago RK. Comparison of various seed priming methods for seed germination, seedling vigour and fruit yield in okra (*Abelmoschus esculentus* L. Moench). Sciantia Horticulturae. 2014;165:75-81.

[101] Janecko A, Dziurka M, Ostrowska A, Biesaga-Koscielniak J, Koscielniak J. Improving vitamin content and nutritional value of legume yield through water and hormonal seed priming. Legume Research. 2015;38:185-193.

[102] Fallahi J, Rezvani-Moghaddam P, Nassiri Mahallati M, Behdani MA, Shajari MA, Amiri MB. 2013. Influence of seed nitrogen content and biofertilizer priming on wheat germination in salinity stress conditions. Archives of Agronomy and Soil Scence. 2013;59:791-891.

[103] Jisha KC. Vijayakumari K, Puthur JT. Seed priming for abiotic stress tolerance: an overview. Acta Physiologia Plantarum. 2013;35:1381-1396. DOI: 10.1007/s11738-012-1186-5

[104] Krol P, Igielski R, Pollmann S, Kepczynska E. Priming of seeds mith methyljasmonate induced resistance to hemibiotroph *Fusarium oxysporum* f. sp. *Lycopersici* in tomato via 12-oxo-phytodienoic acid, salicylic acid, and flavonol accumulation. Journal of Plant Physiology. 2015;179:122-132.

[105] Kalischuk ML, Johnson D, Kawchuk LM. Priming with a double-stranded DNA virus alters *Brassica rapa* seed architecture and facilitates a defense response. Gene. 2015;557:130-137.

[106] Anup CP, Prasad M, Nataraj S, Mayuri NG, Manbali J, Hussain A, Kukkundoor RK. Proteomic analysis of elicitation of downy mildew disease resistance in pearl millet by seed priming with β-aminobutyric acid and *Pseudomonas fluorescens*. Journal of Proteomics. 2015;120:58-74.

[107] Chen K, Fessehaie A, Arora R. Dehydrin metabolism is altered during seed osmopriming and subsequent germination under chilling and desiccation in *Spinacia oleracea* L. cv. Bloomsdale: possible role in stress tolerance. Plant Science. 2012;183:27-36. DOI: 10.1016/j.plantsci.2011.11.002

[108] Maia J, Dekkers BJW, Provart NJ, Ligterink W, Hilhorst HWM. The re-establishement of desiccation tolerance in germinated *Arabidopsis thaliana* seeds and its associated transcriptome. PLoS One. 2011;6:e29123.

[109] Buitink J, Jegger JJ, Guisle I., Wu BL, Wuillème Lamirault G, Bars AL, Meur NL, Becker A, Küster H, Leprince O. Transcriptome profiling uncovers metabolic and regulatory processes occurring during the transition from desiccation sensitive to desiccation-tolerant stages in *Medicago truncatula* seeds. Plant Journal. 2006;47:735-750.

[110] Boudet J, Buitnink J, Hoekstra FA, Rogniaux H, Larré C, Satour P, Leprince O. Comparative analysis of the heat stable proteome of radicles of *Medicago truncatula* seeds during germination identifies late embryogenesis abundant proteins associated with desiccation tolerance. Plant Physiology. 2006;140:1418-1436.

[111] Bewley JD, Bradford KJ, Hilhorst HWM, Nonogaki H. Seeds. Physiology of Development, Germination and Dormancy. 3rd ed. New York, Heidelberg, Dordrecht, London: Springer; 2013. 392 p. DOI: 10.1007/978-1-4614-4693-4

[112] Maurel C. Plant aquaporins: novel functions and regulation properties. FEBS Letters. 2007;581:2227-22236. DOI: 10.1016/j.febslet.2007.03.021

[113] Gomes D, Agasse, A., Thiebaud P, Delrot S, Geros H, Chaumont F. Aquaporins are multifunctional water and solute transporters highly divergent in living organisms. Biochimica Biophysica Acta. 2009;1788:1213-1228. DOI: 10.1016/j.bbamem.2009.03.009

[114] Alleva K, Chara O, Amodeo G. Aquaporins: another piece in the osmotic puzzle. FEBS Letters. 2012;586:2991-2999. DOI:10.1016/j.febslet.2012.06.013

[115] Vander Willigen VC, Postaire O, Tournaire-Roux C, Boursiac Y, Maurel C. Expression and inhibition of aquaporins in germinating *Arabidopsis* seeds. Plant and Cell Physiology. 2006;47:1241-1250. DOI: 10.1093/pcp/pcj094

[116] Liu HY, Yu X, Cui DY, Sun MH, Sun WN, Tang ZC, Su WA. The role of water channel proteins and nitric oxide signaling in rice seed germination. Cell Research. 2007;17:638-649. DOI: 10.1038/cr.2007.34

[117] Ge FW, Tao P, Zhang Y, Wang JB. Characterization of *AQP* gene expressions in *Brassica napus* during seed germination and in response to abiotic stresses. Biologia Plantarum. 2014;58:274-282. DOI: 10.1007/s10535-013-0386-1

[118] Gao YP, Young L, Bonham-Smith P, Gusta LV. Characterization and expression of plasma and tonoplast membrane aquaporins in primed seed of *Brassica napus* during germination under stress conditions. Plant Molecular Biology. 1999;40:635-644. DOI: 10.1023/A:1006212216876

[119] Chen K, Fessehaie A, Arora R. Aquaporin expression during seed osmopriming and post-priming germination in spinach. Biologia Plantarum. 2013;57:193-198.

[120] Kubala S, Wojtyla Ł, Quinet M, Lechowska K, Lutts S, Garnczarska M. Enhanced expression of the proline synthesis gene *P5CSA* in relation to seed osmopriming improvement of *Brassica napus* germination under salinity stress. Journal of Plant Physiology. 2015;183:1-12. DOI: 10.1016/j.jplph.2015.04.009

[121] Schmid, M, Davison TS, Henz SR, Pape UJ, Demar M, Vingron M, Schölkopf B, Weigel D, Lohmann JU. A gene expression map of *Arabidopsis thaliana* development. Nature Genetics. 2005;37:501-506. DOI:10.1038/ng1543

[122] Gattolin S, Sorieul M and Frigerio L. Mapping of tonoplast intrinsic proteins in maturing and germinating *Arabidopsis* seeds reveals dual localization of embryonic TIPs to the tonoplast and plasma membrane. Molecular Plant. 2011;4:180-189. DOI: 10.1093/mp/ssq051

[123] Krishnan P, Joshi DK, Nagarajan S, Moharir AV. Characterisation of germinating and non-viable soybean seeds by nuclear magnetic resonance (NMR) spectroscopy. Seed Science Research. 2004;14:355-362. DOI: 10.1079/SSR2004189

[124] Manz B, Müller K, Kucera B, Volke F, Leubner-Metzger G. Water uptake and distribution in germinating tobacco seeds investigated in vivo by nuclear magnetic resonance imaging. Plant Physiology. 2005;138:1538-1551. DOI: 10.1104/pp.105.061663

[125] Garnczarska M, Tomasz Zalewski T, Kempka M. Water uptake and distribution in germinating lupine seeds studied by magnetic resonance imaging and NMR spectroscopy. Journal of Experimental Botany. 2007;58:3961-3969. DOI: 10.1093/jxb/erm250

[126] Falourd X, Natali F, Peters J, Foucat L. Molecular mobility in *Medicago truncatula* seed during early stage of germination: neutron scattering and NMR investigations. Chemical Physics. 2013;428:181-185. DOI: 10.1016/j.chemphys.2013.10.014

[127] Terskikh VV, Feurtado JA, Ren C, Abrams SR, Kermode AR. Water uptake and oil distribution during imbibition of seeds of western white pine (Pinus *monticola* Dougl. ex D. Don) monitored *in vivo* using magnetic resonance imaging. Planta. 2005;221:17-27. DOI: 10.1007/s00425-004-1426-z

[128] Wojtyla Ł, Garnczarska M, Zalewski T, Bednarski W, Ratajczak L, Jurga S. A comparative study of water distribution, free radical production and activation of antioxidative metabolism in germinating pea seeds. Journal of Plant Physiology. 2006;163:1207-1220. DOI: 10.1016/j.jplph.2006.06.014

[129] Nagarajan S, Pandita VK, Joshi DK, Sinha JP, Modi BS. Characterization of water status in primed seeds of tomato (*Lycopersicon esculentum* Mill.) by sorption properties and NMR relaxation times. Seed Science Research. 2005;15:99-111. DOI: 10.1079/SSR2005200

[130] Pietrzak LN, Frégeau-Reid J, Chatson B, Blackwell B. Observation on water distribution in soybean seed during hydration processes using nuclear magnetic resonance imaging. Canadian Journal of Plant Science. 2002;82:513-519.

[131] Liu Y, Van der Burg WJ, Aartse JW, Van Zwol RA, Jalink H, Bino RJ. X-ray studies on changes in embryo and endosperm morphology during priming and imbibition of tomato seeds. Seed Science Research. 1993;3:171-178. DOI: 10.1017/S0960258500001756

[132] Ashraf M, Foolad MR. Pre-sowing seed treatment: a shotgun approach to improve germination growth and crop yield under saline and non-saline conditions. Advances in Agronomy. 2005;88:223-71. DOI: 10.1016/S0065-2113(05)88006-X

[133] Farooq M, Basra SMA, Wahid A. Priming of field-sown rice seed enhances germination, seedling establishment, allometry and yield. Plant Growth Regulation. 2006;49:285-294. DOI: 10.1007/s10725-006-9138-y

[134] Li X, Zhang L. SA and PEG-induced priming for water stress tolerance in rice seedling. In: Zhu E, Sambath S, editors. Information Technology and Agricultural Engineering. Berlin, Heidelberg: Springer; 2012. 134. p 881-887.

[135] Battaglia M, Covarrubias AA. Late embryogenesis abundant (LEA) proteins in legumes. Frontiers in Plant Science. 2013;4:190. DOI: 10.3389/fpls.2013.00190

[136] Gallardo K, Job C, Groot SPC, Puype M, Demol H, Vandekerckhove J, Job D. Proteomic analysis of Arabidopsis seed germination and priming. Plant Physiology. 2001;126:835-848. DOI: 10.1104/pp.126.2.835

[137] Li F, Wu X, Tsang E, Cutler AJ. Transcriptional profiling of imbibed *Brassica napus* seed. Genomics. 2005;86:718-30. DOI: 10.1016/j.ygeno.2005.07.006

[138] Soeda Y, Konings MCJM, Vorst O, van Houwelingen AMM, Stoopen GM, Maliepaard CA, Kodde J, Bino RJ, Groot SPC, van der Geest AHM. Gene expression programs during *Brassica oleracea* seed maturation, osmopriming, and germination are indicators of progression of the germination process and the stress tolerance level. Plant Physiology. 2005;137:354-368. DOI: 10.1104/pp.104.051664

[139] Oluoch MO, Welbaum GE. Viability and vigor of osmotically primed muskmelon seeds after nine years of storage. Journal of the American Society for Horticultural Science. 1996;121:408-413.

[140] Toorop PE, Van AC, Hilhorst HWM. 1998. Endosperm cap weakening and endo-β-mannanase activity during priming of tomato (*Lycopersicon esculentum* cv. Moneymaker) seeds are initiated upon crossing a threshold water potential. Seed Science Research. 1998;8:483-491. DOI: 10.1017/S0960258500004451

[141] Nawaz J, M, Abdul Jabbar A, Nadeem GA, Sajid M, Subtain MUI, Shabbir I. Seed priming a technique. International Journal of Agriculture and Crop Sciences. 2013;6:1373-1381

[142] de Castro RD, Zheng X, Bergervoet JHW, de Vos CHR, Bino RJ. β-tubulin accumulation and DNA replication in imbibing tomato seeds. Plant Physiology.1995;109:499-504.

[143] Dawidowicz-Grzegorzewska A. Ultrastructure of solid matrix-primed endospermic and nonendospermic seeds. In: Ellis RH, Black M, Murdoch AJ, Hong TD, editors. Basic and Applied Aspects of Seed Biology. Dordrecht: Kluwer Academic Publishers; 1997. p. 479-487.

[144] Dawidowicz-Grzegorzewska A. Ultrastructure of carrot seeds during matriconditioning with Micro-Cel E. Annals of Botany. 1997;79:535-545.

[145] Bolte S, Lanquar V, Soler MN, Beebo A, Satiat-Jeunemaître B, Bouhidel K, Thomine S. Distinct lytic vacuolar compartments are embedded inside the protein storage vacuole of dry and germinating *Arabidopsis thaliana* seeds. Plant and Cell Physiology. 2011;52:1142-1152. DOI: 10.1093/pcp/pcr065

[146] Zheng H, Staehelin A. Protein storage vacuoles are transformed into lytic vacuoles in root meristematic cells of germinating seedlings by multiple, cell type-specific mechanisms. Plant Physiology. 2011;155:2023-2035. DOI: 10.1104/pp.110.170159

[147] Bino RJ, de Vries JN, Kraak L, van Pijlen JG. Flow cytometric determination of nuclear replication stages in tomato seeds during priming and germination. Annual Botany. 1992;69:231-236.

[148] Vázquez-Ramosa JM. de la Paz Sáncheza M. The cell cycle and seed germination. Seed Science Research. 2003;13:113-130. DOI: 10.1079/SSR2003130

[149] Sanchez MDL, Gurusinghe SH, Bradford KJ, Vazquez-Ramos JM. Differential response of PCNA and Cdk-A proteins and associated kinase activities to benzyladenine and abscisic acid during maize seed germination. Journal of Experimental Botany. 2005;56:515-523. DOI: 10.1093/jxb/eri029

[150] Ventura L, Donà M, Macovei A, Carbonera D, Buttafava A, Mondoni A, Rossi G, Balestrazzi A. Understanding the molecular pathways associated with seed vigor. Plant Physiology and Biochemistry. 2012;60:196-206. DOI: 10.1016/j.plaphy.2012.07.031

[151] Bailly C, El-Maarouf-Bouteau H, Corbineau F. From intracellular signaling networks to cell death: the dual role of reactive oxygen species in seed physiology. Comptes Rendus Biologies. 2008;331:806-814. DOI: 10.1016/j.crvi.2008.07.022

[152] Osborne DJ. Biochemical control-systems operating in the early hours of germination. Canadian Journal of Botany. 1983;61:3568-3577. DOI: 10.1139/b83-406

[153] Ashraf M, Bray CM. DNA synthesis in osmoprimed leek (*Allium porrum* L.) seeds and evidence for repair and replication. Seed Science Research. 1993;3:15-23. DOI: 10.1017/S0960258500001525

[154] Law SR, Narsai R, Taylor NL, Delannoy E, Carrie C, Giraud E, Millar AH, Small I, Whelan J. Nucleotide and RNA metabolism prime translational initiation in the earliest events of mitochondrial biogenesis during *Arabidopsis* germination. Plant Physiology. 2012;158:1610-1627. DOI: 10.1104/pp.111.192351

[155] Law SR, Narsai R, Whelan J. Mitochondrial biogenesis in plants during seed germination. Mitochondrion. 2014;19:214-221. DOI: 10.1016/j.mito.2014.04.002

[156] Macovei A, Balestrazzi A, Confalonieri M, Carbonera D. The Tdp1 (tyrosyl-DNA phosphodiesterase) gene family in barrel medic (*Medicago truncatula* Gaertn.): bioinformatic investigation and expression profiles in response to copper- and PEG-mediated stress. Planta. 2010;232:393-407. DOI: 10.1007/s00425-010-1179-9

[157] Macovei A, Balestrazzi A, Confalonieri M, Faé M, Carbonera D. New insights on the barrel medic MtOGG1 and MtFPG functions in relation to oxidative stress response in planta and during seed imbibition. Plant Physiology and Biochemistry. 2011;49:1040-1050. DOI: 10.1016/j.plaphy.2011.05.007

[158] Macovei A, Balestrazzi A, Confalonieri M, Buttafava A, Carbonera D. The TFIIS and TFIIS like genes from *Medicago truncatula* are involved in oxidative stress response. Gene. 2011;470:20-30. DOI: 10.1016/j.gene.2010.09.004

[159] Waterworth WM, Masnavi G, Bhardwaj RM, Jiang Q, Bray CM, West CE. A plant DNA ligase is an important determinant of seed longevity. Plant Journal. 2010;63:848-886. DOI: 10.1111/j.1365-313X.2010.04285.x

[160] Sharma SN, Maheshwari A. Expression patterns of DNA repair genes associated with priming small and large chickpea (*Cicer arietinum*) seeds. Seed Science and Technology. 2015;43:250-261. DOI: 10.15258/sst.2015.43.2.11

[161] Kibinza S, Bazin J, Bailly C, Farrant JM, Corbineau F, El-Maarouf-Bouteau H. Catalase is a key enzyme in seed recovery from ageing during priming. Plant Science. 2011;181:309-315. DOI: 10.1016/j.plantsci.2011.06.003

[162] Kubala S, Wojtyla Ł, Garnczarska M. Seed priming improves salt stress tolerance during germination by modulation of antioxidative capacity. In: Proceedings of the 11th International POG Conference Reactive Oxygen and Nitrogen Species in Plants, Warszawa, 17–19 June 2013, BioTechnologia. 94:223.

[163] Wojtyla Ł, Lechowska K, Kubala S, Garnczarska M. Different modes of hydrogen peroxide action during seed germination. Frontiers Plant Science. 2016;7:66. DOI: 10.3389/fpls.2016.00066

[164] Krasuska U, Gniazdowska A. Nitric oxide and hydrogen cyanide as regulating factors of enzymatic antioxidant system in germinating apple embryos. Acta Physiologiae Plantarum. 2012;34:683-692. DOI: 10.1007/s11738-011-0868-8

[165] Nouman W, Basra SMA, Yasmeen A, Gull T, Hussain SB, Zubair M, Gul R. Seed priming improves the emergence potential, growth and antioxidant system of *Moringa oleifera* under saline conditions. Plant Growth Regulation. 2014;73:267-278. DOI: 10.1007/s10725-014-9887-y

[166] Xu S, Hu J, Li Y, Ma W, Zheng Y, Zhu S. Chilling tolerance in *Nicotiana tabacum* induced by seed priming with putrescine. Plant Growth Regulation. 2011;63:279-290. DOI: 10.1007/s10725-010-9528-z

[167] Islam F, Yasmeen T, Ali S, Ali B, Farooq MA, Gill RA. Priming-induced antioxidative responses in two wheat cultivars under saline stress. Acta Physiologiae Plantarum. 2015;37:153. DOI: 10.1007/s11738-015-1897-5

[168] Farooq M, Basra SMA, Khalid M, Tabassum R, Mahmood T. Nutrient homeostasis, metabolism of reserves, and seedling vigor as affected by seed priming in coarse rice. Canadian Journal of Botany. 2006;84:1196-1202. DOI: 10.1139/b06-088

[169] Sung FJM, Chang YH. Biochemical activities associated with priming of sweetcorn seed to improve vigor. Seed Science and Technology. 1993;21:97-105.

[170] Kaur S, Gupta AK, Kaur N. Effect of osmo- and hydropriming of chickpea seeds on seedling growth and carbohydrate metabolism under water deficit stress. Plant Growth Regulation. 2002;37:17-22. DOI: 10.1023/A:1020310008830

[171] Smith PT, Cobb BG. Physiological and enzymatic characteristic of primed, re-dried, and germinated pepper seeds (*Capsicum annum* L). Seed Science and Technology. 1992;20:503-513.

[172] Job C, Kersulec A, Ravasio L, Chareyre S, Pepin R, Job D. The solubilization of the basic subunit of sugarbeet seed 11-S globulin during priming and early germination. Seed Science Research. 1997;7:225-243. DOI: 10.1017/S0960258500003585

[173] Capron I, Corbineau F, Dacher F, Job C, Côme D, Job D. Sugarbeet seed priming: effects of priming on germination, solubilization of 11-S globulin and accumulation of LEA proteins. Seed Science Research. 2000;10:243-254. DOI: 10.1017/S0960258500000271

[174] Mazor L, Perl M, Negbi M. Changes in some ATP-dependent activities in seeds during treatment with polyethylene glycol and during the redrying process. Journal of Experimental Botany. 1984;35:1119-1127.

[175] Corbineau F, Özbingol, N, Vineland D, Côme D. Improvement of tomato seed germination by osmopriming as related to energy metabolism. In: Black M, Bradford KJ, Vasquez-Ramos J, editors. Seed Biology Advances and Applications: Proceedings of the Sixth International Worshop on Seeds. Merida, Mexico, CABI, Cambridge; 2000. p. 467-474. DOI: 10.1079/9780851994048.0467

[176] Haider S, Pal R. Integrated analysis of transcriptomic and proteomic data. Current Genomics. 2013;14:91-110.

[177] Wong JW, Cagney G. An overview of label-free quantitation methods in proteomics by mass spectrometry. Methods Molecular Biology. 2010;604:273-283.

[178] Pace R, Benincasa P, Ghanem ME, Quinet M, Lutts S. Germination of uintreated and primed seeds in rapeseed (*Brassica napus* var *oleifera* Del.) under salinity and low matric potential. Experimental Agriculture. 2012;48:238-251.

[179] de Lespinay A, Lequeux H, Lambillotte B, Lutts S. Protein synthesis is differentially required for germination in *Poa pratensis* and *Trifolium repens* in the absence or in the presence of cadmium. Plant Growth Regulation. 2010;61:205-214.

[180] de Lespinay A. Study of seed priming mechanisms of three plant species used in revegetation of industrial sites. PhD Thesis, Université catholique de Louvain, Belgium. 2009. 253 p.

[181] Bonsager BC, Finnie C, Roepstorff P, Svensson B. Spatio-temporal changes in germination and radicle elongation of barley seeds tracked by proteome analysis of dissected embryo, aleurone layer, and endosperm tissues. Proteomics. 2007;7:4528-4540.

[182] Catusse J, Meinhard J, Job C, Strub JM, Fischer U, Pestova E, Westohoff P, Van Dorselaer A, Job D. Proteomics reveals potential boiomarkers of seed vigor in sugarbeet. Proteomics. 2011;11:1569-1580.

[183] Zhang H, Wang WQ, Liu SJ, Moller JM, Song SQ. Proteome analysis of poplar seed vigor. PLoS One. 2015;10:e0132509.

[184] Rajjou L, Belghazi M, Huguet R, Robin C, Moreau A, Job C, Job D. Proteomic investigation of the effect of salicylic acid an Arabidopsis seed germination and establishment of early defense mechanisms. Plant Physiology. 2006;141:910-923.

[185] Tanou G, Job C, Rajjou L, Arc E, Belghazi M, Diamentidis G, Molasiotis A, Job D. Proteomics reveals the overlapping roles of hydrogen peroxide and nitric oxide in the acclimation of citrus plants to salinity. Plant Journal. 2009;60:795-804.

[186] Wu X, Li N, Li H, Tang H. An optimized method for NMR-based plant seed metabolomics analysis with maximized polar metabolite extraction efficiency, signal-to-noise ratio, and chemical shift consistency. Analyst. 2014;139:1769-1778.

[187] Posmyk MM, Corbineau F, Vinel D, Bailly C, Côme D. Osmoconditioning reduces physiological and biochemical damage induced by chilling in soybean seeds. Physiologia Plantarum. 2001;111:473-482.

[188] Tallowin JRB, Rook AJ, Brookman SKE. The effect of osmoting resowing treatment on laboratory germination in a range of wild flower species. Annals of Applied Biology. 1994;124:363-370.

[189] Wright B, Rowse H, Whipps JM. Microbial population dynamics on seeds during drum and steeping priming. Plant and Soil. 2003;255:631-640.

[190] Chiu KY, Chen CL, Sung JM. Effects of priming temperature on storability of primed sh-2 sweet corn seeds. Crop Science. 2002;42:1996-2003.

[191] Hill HJ, Cunningham JD, Bradford KJ, Taylor AG. Primed lettuce seeds exhibit increased sensitivity to moisture content during controlled deterioration. HortScience. 2007;42:1436-1439.

[192] Schwember AR, Bradford KJ. Drying rate following priming affect temperature sensitivity of germination and longevity of lettuce seeds. HortScience. 2005;40:778-781.

[193] Hacisalihoglu G, Taylor AG, Paine DH, Hildebrand MB, Khan AA. Embryo elongation and germination rate as sensitive indicators of lettuce seed quality: priming and ageing studies. HortScience. 1999;34:1240-1243.

[194] Butler LH, Hay FR, Ellis RH, Smith RD, Murray TB. Priming and redrying improve the survival of mature seeds of *Digitalis purpurea* during storage. Annals of Botany. 2009;103;1261-1270.

[195] Hussain S, Zheng M, Khan F, Khaliq A, Fahad S, Peng S, Huang J, Cui K, Nie L. Benefits of rice seed priming are offset permanently by prolonged storage and the storage conditions. Scientific Reports. 2015;5:8101.

[196] Buitink J, Hemminga MA, Hoekstra FA. Is there a role for oligosaccharides in seed longevity? An assessment of intercellular glass stability. Plant Physiology. 2000;122:1217-1224.

Chapter 2

Recent Advances in Seed Enhancements

Irfan Afzal, Hafeez Ur Rehman,
Muhammad Naveed and
Shahzad Maqsood Ahmed Basra

Additional information is available at the end of the chapter

http://dx.doi.org/10.5772/64791

Abstract

Seed quality is vital to sustainable crop production and food security. Seed enhancements include physical, physiological and biological treatments to overcome germination constraints by uniform stands, earlier crop development and better yields. Improved germination rates and seedling vigour are due to reduced emergence time by earlier start of metabolic activities of hydrolytic enzymes and resource mobilization. Nutrient homeostasis, ion uptake, hormonal regulation, activation of antioxidant defence system, reduced lipid peroxidation and accumulation of compatible solutes are some mechanisms conferring biotic and abiotic stress tolerance. Several transcription factors for aquaporins, imbibitions, osmotic adjustment, antioxidant defence and phenylpropanoid pathway have been identified. However, the knowledge of molecular pathways elucidating mode of action of these effects, reduced longevity of primed or other physical and biological agents for seed treatments and market availability of high-quality seeds are some of the challenges for scientists and seed industry. In this scenario, there is need to minimize the factors associated with reduced vigour during seed production, improve seed storage and handling, develop high-tech seeds by seed industry at appropriate rates and integrate agronomic, physiological and molecular seed research for the effective regulation of high-quality seed delivery over next generations.

Keywords: seed priming, biopriming, coating, magnetic seed stimulation, seed vigour

1. Introduction

Good-quality seed has a significant potential of increasing on-farm productivity and enhancing food security [1]. Seed quality is the foundation for profitable production and marketing [2, 3]. High-quality seeds are genetically and physically pure, vigorous and free from insect pests and pathogens [4]. High-quality seeds with enhanced vigour contribute nearly 30% of the total production. Plant uniformity is an expression of high seed quality achieved by high vigour of seeds [5]. Seed quality is influenced by several factors during seed development, such as maturation, harvesting, drying, cleaning, grading, packing and storage. Farmers and growers are constantly looking for high-quality seeds to ensure uniform field establishment and increased production [6].

Availability, quality and cost of seeds influence the global production and ultimately food security [7]. The informal seed systems (farmers organized and managed without legal documentation) constitute for 75–90% of their food crop cultivation [8]. In the developing world, informal seed systems remain the prevailing source of seed for smallholder farmers. Improper storage environment, sensitivity of germinating seeds and young seedlings to dehydration stress lead to loss of desiccation tolerance with seed hydration [9–11] and predicted climate change (erratic rainfall patterns and unpredictable temperature extremes) may further exacerbate seed quality. Low-vigour seeds can be improved using a variety of seed technologies that will thrive under small holder cultivation conditions and also improve the supply of good-quality seed in the local seed industry.

Efficient seed germination and early seedling establishment are important for commercial agriculture because they represent the most susceptible stages of the life cycle of crop plants [12]. Rapid and uniform seedling emergence leads to successful establishment as it produces a deep root system before the upper layers of soil dry out, harden, or reach supra-optimal temperatures [13]. Germination begins with water uptake by seed and ends with the emergence of the embryonic axis, usually the radicle [14]. A wide range of techniques are now used to help sowing seeds and to improve or protect seedling establishment and growth under the changing environments and seedbed constraints. These techniques constitute the postharvest processing necessary to prepare seed for sowing and optional treatments that are generally described in the industry and scientific literature as 'seed enhancements' or 'seed treatments'. Many scientists have suggested techniques for improving crop germination performance in the field keeping in view the responses of seed to temperature and water availability in the soil. These techniques may be differentiated into physiological (seed priming, coating and pelleting), physical (magnetic, radiation and plasma) and biological (seed enhancements) aspects [15–20]. In 2015, the projected value by global chemical seed treatment industry was up to $ 5.4 billion. Bayer Crop Sciences and Syngenta have 75% share in seed treatment market. In this chapter, we will focus on physiological, biological and physical enhancements of seeds.

Several reports are available, for instance, Heydecker and Coolbear [15] had reported on seed treatments to break dormancy, improve germination and impart stress tolerance and subsequently Taylor et al. [17] continued this work. Halmer [18, 21] focused on practical aspects of seed treatment technologies and categorized it into conditioning, protection and physiological

http://dx.doi.org/10.5772/64791

enhancements. Bray [22] and McDonald [23] further continued work on exploring the mechanisms of physiological enhancements especially seed priming. Regarding physical enhancement, only one key review [24] addressed the effect of magnetic field on growth and yield of crops without primarily focusing on seed germination. Up to now, 1253 research articles have been published on seed priming and out of that almost 100 articles are being published every year since 2010. While there is no consistency in publications on magnetic field treatments, i.e., on this topic roughly 3–4 publications per year and a total of 164 articles have been published up to now [25]. With the recent advances in molecular biology of seeds, here we presented conceptual insights into physiological, biochemical, morphological and biophysical markers that can be used for further improvement of seed quality of crop plants. In addition, in-depth mechanisms of seed germination promotion by physiological, biological and physical seed treatments have also been discussed.

2. Seed enhancements

Seed enhancements or seed invigoration are the post-harvest treatments used for improving the germination and growth of seedlings required at the time of sowing [17]. Many shotgun approaches are being used for seed enhancement for the last 24 years, which includes seed priming, magnetic stimulation, seed pelleting and coating [17, 26, 27].

2.1. Seed enhancement using physical agents

Physical treatments are applied externally without any hydration or application of chemical materials to the seeds. The main purpose is to enhance germination and seedling establishment. The mechanism of seed invigoration with physical techniques is still unknown. The work on exposure of seeds to radiation was started in early 1980s and now a number of studies have been focused on the use of plasma technology for seed invigoration of agronomic and horticultural crops. Magnetic field treatments are being considered as effective seed enhancement tools for agronomic and horticultural crops; however, their application is limited at large scale. Among physical methods, magnetic field and irradiation with microwaves or ionizing radiations are the most promising pre-sowing seed treatments [28]. Thus, physical seed enhancements are an alternative approach to other chemical seed invigoration treatments, which provide better solution for the growing world seed market.

2.1.1. Magnetic fields for seed treatments

Magnetic seed stimulation involves identifying the magnetic exposure dose to affect the germination, early seedling growth and subsequent yield of crop plants [29]. The magnetic exposure dose is the product of the flux density of magnetic field and of the time to exposure. The flux density of magnetic field varies with static or alternating magnetic fields exposure to seeds. Magnetic field ensures the quick germination, uniform crop stand establishment and yield of many agronomic and horticultural crops [30, 31]. These not only increase the rate of germination, growth and yield [32] but also reduce the attack of pathogenic diseases [33, 34].

Magnetic field exposure increases the germination of non-standard seeds and also improves their quality. Magnetic field influences the initial growth stage of the plants after the germination [35]. In recent years, work on magnetic-treated water revealed that plant growth and seed germination were improved by priming [36].

2.1.2. *Plasma seed treatments*

Plasma application in agriculture and medicine is a recent advancement [37–40]. The agricultural aspects include seed germination and plant growth. Many researches report that germination and growth enhancement mechanism is affected by use of plasmas with several gases as aniline, cyclohexane and helium [41, 42]. To enhance seed development and plant-growth microwave plasma, magnetized plasma and atmospheric plasma are adopted treatments [43, 44]. The effect of gases is much commonly studied in plasmas treatments. Various reports revealed that the quality of plant development controlling thiol groups is diversified by redox reaction persuaded by the active oxygen species of water vapour plasma [45].

Non-thermal plasma radiations are applied in agriculture as alternative to scarification, stratification and priming helped to improve the plant growth [46]. Plasma helps to attain zero seed destruction, no chemical use and environment friendly treatments to seeds [41, 46, 47]. Plasma treatment improves seed quality and plant growth [43, 48]. Seed exposure to plasma also resulted in alterations of enzymatic activity [45] and caused sterilization of seed surface [47].

Plasma chemistry can tune seed germination by delaying or boosting with application of plasma-treated deposits on seed surfaces [41]. The recent important plasma-related investigation includes the practice of microwave discharges [43] and low-density radio frequency (RF) discharges [49, 50]. The discharge of atmospheric pressure and the discharge of coplanar barrier have been assessed in recent studies [41, 48]. The investigation of various seed germination patterns was implemented on different seeds including wheat, maize, radish, oat, safflower and blue lupine [43, 46, 48, 50]. Safflower seeds expressed 50% greater germination rate when treated with radio frequency plasma for 130 min with argon [46]. Soybean seeds were treated with cold plasma treatment with 0, 60, 80, 100 and 120 W for 15 s and found positive effects of cold plasma treatments on seed germination and seedling growth of soybean [51].

2.1.3. *Radiation seed treatments*

With recent advancements in agriculture, gamma radiations can improve plant characteristics such as precocity, salinity tolerance, grain yield and product quality in suboptimal environment depending upon the level of irradiation [52]. Second, gamma radiation can also sterilize agricultural products to prevent pathogen infestation thus increasing conservation time during storage and trading [53].

The biological effects of radiations is based on chemical interaction with biomolecules and water to produce free radicals that can manipulate biomolecules and induce cell to switch on antioxidant system [54] that prepared the defensive shield against upcoming stresses [55, 56].

In spite of the conventional seed enhancements, physics has manipulated radiation dose to trigger biochemical reactions necessary for seed germination without affecting seed structural integrity and collateral DNA damage [57]. It was found that the low dose of gamma radiation (up to 20 Gy) on germination of three varieties of Chinese cabbage shows a positive impact [58].

2.2. Physiological seed enhancements

2.2.1. *Seed priming*

Seed priming is a pre-sowing approach for influencing the seedling development by stimulating pre-germination metabolic activities prior to the emergence of radicle and improvement in the germination rate and performance of plant [16, 17]. Seed priming is a controlled hydration process in which seeds are dipped in water or any solution for a specific time period to allow the seed to complete its metabolic activities before sowing and then re-dried to original weight [15, 16].

Priming treatments include osmopriming by polyethylene glycol (PEG) or a salt solution [59], hydropriming [16, 60], solid matrix priming in which seeds are soaked in inert medium of known matrix potential [63] and hormonal priming [62]. A balance of water potential between osmotic medium and seed is necessary for conditioning, and different non-penetrating agents such as organic solutes and salts are used for this purpose [63]. Furthermore, these priming treatments show positive response only at sub-optimal or supra-optimal field conditions such as drought [64], excessively high or low temperatures [60, 65] and salinity [59].

2.2.1.1. *Hydropriming*

Hydropriming is a controlled hydration process that involves seed soaking in simple water and then re-drying to their initial moisture [59, 63]. No chemical is used during this technique but some cases of non-uniform hydration causes uneven germination [66]. Among the different seed enhancement techniques, hydropriming could be a suitable treatment under salinity stress and drought-prone environments [67].

Hydropriming as a risk free, simple and cheap technique has become popular among farmers, with promising effects in the context of extensive farming system [68]. Hydroprimed seeds produced healthy seedlings, which resulted in uniform crop stand, drought resistance, early maturity and somewhat improved yield.

2.2.1.2. *Osmopriming*

Osmopriming involves seed hydration in an osmotic solution of low water potential such as polyethylene glycol or a salt solution under controlled aerated conditions to permit imbibition but prevent radical protrusion [67]. For osmopriming, mostly polyethylene glycol or salt solution is used to regulate water uptake and to check radicle protrusion [64]. Most commonly used salts for osmopriming are potassium chloride (KCl), potassium nitrate (KNO_3), sodium chloride (NaCl), magnesium sulphate ($MgSO_4$), potassium phosphate (K_3PO_4), calcium chloride ($CaCl_2$) and potassium hydrophosphate (KH_2PO_4). All these salts provide nutrient like nitrogen to the germinating seed, which is required for the protein synthesis during the germination process. However, these salts rarely cause nutrient toxicity to the germinating young seedlings [63]. Osmopriming induced more rapid and uniform germination and resulted in decreased mean germination time.

2.2.1.3. *Hormonal priming*

Plant-growth hormones or their derivatives contained by several products are indole-3-butyric acid (IBA), an auxin and kinetin type of cytokinin. Cytokinins play a vital role in all phases of plant development starting from seed germination up to senescence [70]. Priming with optimum concentration of cytokinins has been reported to increase germination, growth and yield of many crop species [16]. Gibberellic acid (GA_3) is known to break seed dormancy, enhance germination, hypocotyl growth, internodal length, and cell division in the cambial zone and increase the size of leaves. GA has stimulatory effect on hydrolytic enzymes, which speed up the germination and promote seedling elongation by degrading the cells surrounding the radicle in cereal seeds [69, 71].

Various naturally occurring growth promoting substances such as moringa leaf extract, chitosan, sorghum water extract and seed weed extract [62, 65] are commonly used for seed priming. Moringa (*Moringa oleifera* L.) as a natural source of plant-growth regulators contains cytokinins as zeatin [72]. In addition, moringa leaf extracts contain higher concentrations of various growth enhancers such as ascorbates, phenolic compounds, K, and Ca. Priming maize seed with moringa leaf extract reduces mean germination (MGT) and T_{50} with increased germination index and germination count that ultimately improved seedling growth by increasing chlorophyll content, amylase activity and total sugar contents under chilling conditions [62]. Moringa leaf extract diluted up to 1:36 with water was applied on various field crops and 35% increase in the yield of sugarcane, sorghum, maize, turnip and bell pepper was observed [72]. Nonetheless, moringa leaf extracts being low cost can be a viable option for improving the productivity of resource poor farmers.

2.2.1.4. *Nutrient priming*

The application of micronutrients with priming can improve stand establishment, growth and yield; furthermore, the enrichment of grain with micronutrients is also reported in most cases [73]. Many researchers proved the potential of nutrient priming in improving wheat, rice and forage legumes. Among micronutrients, Zn, B, Mo, Mn, Cu and Co are highly used as seed treatments for most of the field crops [74–76].

Seed treatment with micronutrient is a potentially low-cost way to improve nutrition of crops. Farmers have responded in South Asia in a positive way in the seed treatment, which is a simple technique soaking seeds in water overnight before planting [77]. Seed priming with zinc salts is used to increase growth and disease resistance of seedlings.

2.3. Biological seed enhancements

2.3.1. *Bacterial seed agents*

Plant-growth-promoting rhizobacteria (PGPR) are free-living, soil-borne bacteria, which when applied to soil, seeds or roots promote the growth of the plant or reduce the incidence of diseases from soil-borne plant pathogens. PGPR can influence plant growth either directly or indirectly through fixation of atmospheric nitrogen, solubilization of phosphorus and zinc and producing siderophores, which can solubilize/sequester iron, synthesize phytohormones, including auxins, cytokinins and gibberellins to stimulate plant growth, and synthesize ACC-deaminase enzyme by modulation of ethylene level under stress conditions [78, 79].

Among various genera of PGPR endophytes are good priming agents because they colonize roots and create a favourable environment to develop and function with their hosts—symbiotic partner.

Biopriming is a new technique of seed enhancement integrating biological (inoculation of seed with beneficial organism to protect seed) and physiological aspects (seed hydration) to promote plant growth, development and suppression of diseases. It is used as an alternative approach for controlling many seed- and soil-borne pathogens. Seed priming with beneficial microorganisms (bacteria and fungus) often result in more rapid growth and increase plant vigour and may be useful under adverse soil conditions. Besides diseases control, the application of PGPR as a biopriming agent for biofertilization is an attractive option to reduce the use of chemical fertilizers [80, 81]. PGPR that have been tested as co-inoculants with rhizobia include strains of the following rhizobacteria: *Azotobacter* [82], *Azospirillum* [83], *Bacillus* [84], *Pseudomonas* [85, 86], *Serratia* [86] and *Streptomyces* [87].

2.3.1.1. *Role of a bacterial biopriming agent in plant-growth promotion*

The Plant growth promoting bacteria (PGPB) are a heterogeneous group of beneficial microorganisms present in the rhizosphere, on the root surface or inside plant tissues, and are able to accelerate the growth of plants and protect them from biotic and abiotic stresses [88–90]. Beneficial effects of biopriming have been reported in several vegetable seeds [91]. Priming of tomato seed with beneficial bacteria improved the rate of germination, seedling emergence and growth of plant [92]. The beneficial response of biopriming on seed germination and seedling vigour in chilli was reported [93]. Similarly, improvement in okra growth and yield was reported up to 60% when seeds were bioprimed with *P. fluorescens* culture [94]. In experiments where lettuce plants were treated with *Bacillus* strains, it was observed that after two weeks the tissues of roots and shoots contained a greater amount of cytokinin than control plants [95, 96]. The accumulation of cytokinins was associated with a 30% increase in plant biomass

2.3.1.2. *Role of a bacterial biopriming agent in plant disease control*

Seed enhancement by biopriming agents involves coating/soaking the seed with one biological agent or microbial consortium, then incubating the seed under optimum (temperature, moisture) conditions.

Bacterial strain	Target plant	Condition	Proposed mechanism	Effects	References
Rhizobium leguminosarum bv. *Viciae*	Faba bean (*Vicia faba*)	Green house and field	Improved nitrogenase activity and production of IAA	Increased nodulation and nitrogen fixation under drought and salinity stress	[19]
Pseudomonas spp. NUU1 and *P. fluorescens* NUU2	Wheat (*Triticum aestivum*)	Pot experiment	Auxin production	Stimulated the shoot and root length and dry weight	[149]
Pseudomonas fluorescens MSP-393	Rice (*Oryza sativa*)	Green house	Production of osmolytes	Increased plant growth and vigour	[150]
Pseudomonas putida GAP-P45	Sunflower (*Helianthus annuus*)	Field experiment	Production of exopoly saccharides, biofilm	increased the survival, plant biomass, and root adhering soil/ root tissue ratio of sunflower seedlings under drought stress	[151]
Rhizobium and Pseudomonas species	Maize (*Zea mays*)	Pot experiment	decreases in osmotic potential, and increase in osmoregulant (proline) production, maintenance of relative and selective uptake of K ions.	Promote plant growth and increased relative water contents	[152]
Pseudomonas chlororaphis isolate TSAU13	Cucumber (*Cucumis sativus*) and Tomato (*Solanum Lycopersicum*)	Green house	Antibiosis	Stimulated shoot growth, dry matter and the fruit yield of tomato and cucumbers under saline conditions	[153]
T. Harzianum T22 *Rifai* KRL-AG2	Onion (*Allium cepa* L.)	Axenic trial	Osmotic adjustment through physiological responses	Increased germination %age, shoot length and seedling fresh weight under saline conditions	[20]
Piriformospora indica	Chinese cabbage (*Brassica rapa* subsp. *pekinensis*)	Pot experiment	Involved in expression of diverse stress-related genes	Promotes root and shoot growth, and promotes lateral root formation	[154]
Neotyphodium	Arizona fescue (Festuca arizonica Vasey)	Green house	Regulate stomatal conductance	Increased relative growth rates, High W.U.E and biomass yield under drought	[155]

Table 1. Observed effects of plant-beneficial bacteria in regard to plant-growth promotion and stress tolerance.

Biopriming of seeds with different bacterial strains particularly rhizobacteria have been shown to be effective in suppressing disease infection by inducing a resistance mechanism called 'induced systemic resistance' (ISR) in varied agronomic and horticultural crops [97]. Among various bacterial genera, *Bacillus* and *Pseudomonas* spp. are ubiquitous rhizosphere inhabitant bacteria that are the most studied biopriming agents reported as disease suppressing in plants [98]. Priming seeds of many crops with biological control agents (BCA), *Bacillus subtillus* and *Pseudomonas fluorescens* are the most effective approach for controlling seed and root rot pathogens [99, 100] and as a substitute for chemical fungicides without any risk to human, animal and the environment.

2.3.1.3. Seed enhancement by alleviating abiotic stresses using biopriming

Seed priming with beneficial microorganisms may promote plant growth and increases abiotic stress tolerance in arid or semiarid areas [101]. PGPB are adapted to adverse conditions and protect plants from the deleterious effects of these environmental stresses, thus increasing crop productivity [102]. Bioprimed seeds with *Enterobacter* sp. P-39 showed maximum improvement in germination and seedling growth of tomato under osmotic stress [103]. **Table 1** shows the selected examples of beneficial response of biological inoculants for enhancing growth and yield of various crops under normal and stress conditions.

2.3.2. Fungal seed agents for biopriming

In this approach, beneficial bacterial and fungal agents are exploited for the purpose of biopriming of seeds to enhance growth, yield and mitigation of biotic and abiotic stresses. It is an environmental friendly, socially accepted approach and also offers an alternative to the chemical treatment methods gaining importance in seed, plant and soil health systems. Seed biopriming enhanced drought tolerance of wheat as drought-induced changes like photosynthetic parameters and redox states were significantly improved by *Trichoderma* sp. under stress conditions over control. Very recently, Junges et al. [84] compared the potential of biopriming (*Trichoderma* and *Bacillus* spp.) with commercial available products Agrotrich plus® and Rhizoliptus® for enhancing growth and yield of beans. Results revealed that biopriming with spore or bacterial cell suspensions promoted bean seedling growth compared to other techniques.

2.4. Seed coating and pelleting

Seed film coating, pelleting, priming and inoculation are globally practiced seed treatments [104] used with the objectives of enhancing plantability, distribution, germination and storage of seeds. These techniques aim to apply adhesive films, fungicides, herbicides, growth promoters and biological agents [3, 91, 105]. Seed coating is carrier of chemical materials to support seedling growth [106]. Compounds such as growth regulators, inoculants, micronutrients, fungicides, insecticides and other seed protectants are applied to the pellet to enhance seed performance [107].

Seed coating demands uniform application of inert material over the seed surface. This also helps to protect the seed from soil and seed-borne pathogens [17]. Pharmaceutical industry uses seed polymer coating for a constant application of numerous materials to seeds. The commercially available plasticizers, polymers and colourants (commercially they are readily available to be used as liquid) are applied as film formulations [108]. However, the exact composition of coating material is a carefully guarded secret by the companies who develop them. Usually, coating material contains binders, fillers (e.g., polyvinyl alcohol, gypsum and clay) and an intermediate layer (e.g., clay, polyvinyl acetate and vermiculite). Seed agglomeration is an alternate coating technology with the purpose to sow multiple seeds of the same seed lot, or multiple seeds of different seed lots, varieties or species [109].

3. Mechanisms of seed enhancements

3.1. Physiological and biochemical aspects

Improved crop performance through pre-sowing treatments depends on the nature of compounds used for priming and their accumulation under abiotic stresses. These compounds include inorganic salts, osmolytes, phytohormones, tertiary amino compounds such as glycinebetaine, amino acids and sugar alcohols including bioactive compounds from micro-organisms. For instance, the application of compatible solutes as seed priming improves salinity resistance by cytosolic osmotic adjustment indirectly by enhancing regulatory functions of osmoprotectants [110, 111]. Chilling-induced cross-adaptation salt tolerance in wheat is associated with enhanced accumulation of beneficial mineral elements (K^+ and Ca^{2+}) in the roots and reduced uptake of toxic Na^+ in the shoots through ionic homeostasis and hormonal balance with greater concentrations of indoleacetic acid, abscisic acid, salicylic acid and spermine in chilled wheat seeds [112]. In flooded soils, improved stand establishment in rice through seed priming is related to enhanced capacity of superoxide dismutase (SOD) and catalase (CAT) activities to detoxify the reactive oxygen species in seeds and greater carbohydrate mobilization. These effects are more pronounced in tolerant genotypes that emphasize to combine crop genetic tolerance with appropriate seed treatments to improve seedling establishment of rice sown in flooded soils [113].

Such enhanced remobilization efficiency in seed embryos of cereals coated with hydro-absrobers is related to change in activities of enzymes for sucrose breakdown upon moisture absorption. Coated seeds absorb more moisture that creates anoxic conditions in developing embryos but genetic difference are found for sucrose breakdown in rye, barley and wheat with change in invertase activities due to difference in timing of imbibitions [114].

Beneficial effects of magnetic seed stimulation are associated with various biochemical, cellular and molecular events [115]. Pre-sowing magnetic seed treatment also increases ascorbic acid contents [33] by stimulating the activity of the enzymes and proteins [116]. Physiological and biochemical properties also increase due to enhanced metabolic pathway by the free movement of ions [117]. However, its biochemical and physiological mechanisms are still poorly understood [118].

3.2. Molecular aspects

Favourable effects of priming at cellular level include RNA and protein synthesis [22]. Seed priming induces several biochemical changes within the seed needed for breaking seed dormancy, water imbibition, enzymes activation, hydrolysis of food reserves and mobilization of inhibitors [119]. At cellular level priming initiates cell division transportation of storage protein [120]. Higher germination rate and uniform emergence of primed seed is due to metabolic repair with increased production of metabolite required for the germination [121, 122] during the imbibition process. Priming increased the production and activity of α-amylase within germinating seeds, thus increased the seed vigour [65].

Several proteins and their precursors for regulation involved in different steps of seed germination or priming have been identified using model plant Arabidopsis. The expression of these proteins such as actin isoform or a WD-40 repeat protein occurs in imbibition and cytosolic glyceraldehyde-3-phosphate dehydrogenase in the seed dehydration process [123]. Priming-induced changes in proteins levels have been identified as peroxiredoxin-5, 1-Cys peroxiredoxin, embryonic protein DC-8, cupin, globulin-1 and late embryogenesis abundant protein. The expression of these proteins led to improved seed germination and the expression of these embryo proteins remained unchanged even after priming [124].

A major quantitative trait locus (QTL) Htg6.1 of seed germination responsive to priming under high temperature stress using a recombinant inbred line (RIL) of lettuce (*Lactuca sativa* L.) has been identified. The expression of this QTL at high temperature is coded by a gene *LsNCED4* encoding the key enzyme, i.e., 9-cis-epoxycarotenoid dioxygenase, of the abscisic acid biosynthetic pathway and maps precisely with Htg6.1. However, *LsNCED4* gene expression was higher in non-primed seeds after 24 h of imbibition at high temperature compared to the expression of *LsGA3ox1* and *LsACS1* genes encoding enzymes of gibberellins and ethylene biosynthetic pathways, respectively. *LsNCED4* gene expression was reduced after priming and when imbibition was carried out at the same temperature . The seed response to priming in terms of germination and temperature sensitivity is associated with temperature regulation of hormonal biosynthetic pathways [125].

Osmopriming induced quantitative expression of stress-responsive genes such as CaWRKY30, PROX1, Osmotin for osmotic adjustment, Cu/Zn SOD for antioxidant defence and CAH for phenylpropanoid pathway. The same genes were induced earlier or at higher levels in response to thiourea priming at low temperature. The expression of these genes imparts cold tolerance in capsicum seedlings [126]. Notably, high levels of other plant-growth hormones, such as indolyl-3-acetic acid (IAA) and abscisic acid (ABA), were also observed. The authors suggested that *Bacillus* strains have dual effect on plant-growth promotion and accumulation of cytokinins by increasing other routes of synthesis of hormones such as IAA and ABA, as well as interfering in other hormonal balance synthesis such as gibberellins (GA).

Using advanced molecular tools such as proteomics may help to detect protein markers that can be used to unravel complex development process of seed vigour of commercial seed lots, or analysis of protein changes occur in industrial seed priming treatments to accelerate seed germination and improve seedling uniformity.

4. Seed enhancements and plant development

4.1. Modulation of seedling growth

Seedling vigour is important to help ensuring good crop establishment. Pre-sowing seed treatments offer pragmatic solution to poor seedling establishment by overcoming the germination constraints under normal and adverse conditions. Several researches have shown the potential of chemical priming, use of macro- and micronutrients, natural compounds of plant origin and plant-growth-promoting bacteria including water under greenhouse and field conditions. Most of the priming techniques such as osmopriming and on-farm priming have been optimized for specific crops for soaking duration and concentration. For chemical priming, polyamines including spermine, spermeidine and putrescine, calcium chloride ($CaCl_2$), potassium chloride (KCl), NaCl, KH_2PO_4, KNO_3, PEG, hydro-absorbers such as humic acid and biplantol for seed coating and naturally occurring molecules such as nitric oxide (NO), hydrogen sulphide (H_2S), H_2O_2, ascorbate, salicylic acid, indoleamine molecule melatonin (Mel) and most recently growth promoting cytokinin-rich moringa leaf extracts are commonly being evaluated. The endogenous levels of naturally occurring molecules when applied as seed priming may increase initially and later with subsequent improved growth.

The beneficial effects of seed priming have been documented in cereals, sugar crops, oilseeds and horticultural crops. Early seedling growth by pre-sowing seed treatments is due to improved germination rate, reduced time of germination or emergence, and uniform and enhanced germination percentage contributed by enhanced mobilization of germination metabolites from endosperm towards growing embryonic axis. However, variation in germination rates with seed coating thickness and composition has been found which ultimately affects the mobilization efficiency of seed reserves. Therefore, the use of hydro-absorbers is suggested for coated seeds to enhance the efficiency of germination metabolites which may differ among species [114].

Seed priming with nutrients usually increases the seed contents of primed nutrients, which may be translocated to the growing seedling to support the seedling development [127]. Improved seedling growth and dry mass may be attributed to enhanced nutrient uptake and enzymes associated under deficient conditions and offer perspective for improved seed quality at crop harvesting [128]. Priming mediated by manganese (Mn) has also significant effect on the growth and yield performance of crops. In comparison to soil application, Mn priming improved stand establishment, growth, yield and grain contents [129]. Another researcher also noticed that priming with cobalt nitrate had increased growth attributes and subsequent yield of pigeon pea [130].

The concentration of these nutrients may be toxic when used in relatively higher concentration. For instance, priming with 0.5% Boron solution completely suppressed the germination and growth in rice [27] and 0.1 M $ZnCl_2$ and 0.5 M $ZnSO_4$ in wheat [131]. Seed priming induced early vigour indices have been associated with suppression of weeds in primed stand of aerobic rice [132]. Germination, shoot biomass and total root length were increased in seeds of cultivar IR74 containing Pup1 QTL after water priming. This suggests that seed management approaches may be combined with genetics to improve the crop establishment in different crops including rice under P-deficient conditions [133].

Pre-sowing magnetic seed treatment of wheat seeds has an effect on the germination, and the growth rate was increased to 23% while the germination rate was 100% in the laboratory and less time was taken with 15 min treatment [134].

4.2. Effects on crop phenology

Plants grown by primed seeds usually emerge faster and complete other developmental stages such as tillering, flowering and physiological maturity earlier than seeds without priming [27, 73, 107, 131]. This developmental plasticity of priming may be beneficial when crop planting is delayed due to adverse climatic conditions such as low temperature or high rainfall at sowing, high temperature at reproductive stage and may help plant to avoiding detrimental conditions by earlier maturity [135] without yield decrease. In fact, earlier and vigorous crop stand usually captures more resources of water and nutrients through better root system and had larger leaf area and duration with enhanced photo-assimilation that subsequently contributes towards better yield [73, 131]. However, integrated studies combining seed priming with other crop husbandry practices such as planting geometry, irrigation and fertilization may be interesting in crop stress and nutrient management for improved resource use efficiency.

4.3. Yield improvement

Seed priming benefits are not usually end up with improved crop stand. Several studies report long-lasting effects on yield-associated advantages in terms of increased growth rates, high dry matter production and produce quality by improving crop resistance to biotic and abiotic stresses. A very few reports showing no yield improvement by seed priming are available [136, 137]. Seed priming improved yield is due to reduced weed biomass, higher leaf area index and panicles/m^2 in aerobic and submerged rice, respectively [132, 138, 139], improved crop nutritional status of nutrients primed in maize under low temperature stress [127], comparatively better dry matter production with higher tissue Zn concentration with Zn seed priming in rice [140], reduced spikelet sterility in direct seeded rice irrigated with alternate wetting and drying (AWD) [131] and under system of rice intensification (SRI) condition with improved crop growth and higher tillering emergence [141]. Likewise, early planting spring maize stimulated seedling growth due to increased leaf area index, crop growth and net assimilation rates, and maintenance of green leaf area at maturity [135], better stand establishment in no tilled wheat under rice-wheat system [142], with enhanced tillering emergence and panicle fertility and with B nutrition under water saving rice cultivation [143], GA_3 priming induced modulation of ions uptake (Na^+, K^+) and hormonal homeostasis under salinity in wheat [144], in combination with gypsum + FYM treatment by ameliorating effects on plant growth [145] and improving performance of poor quality wheat seeds under drought stress [146]. Nonetheless, another researcher observed improved yield due to stand establishment and increasing panicle number by coating rice seeds with Zn-EDTA or ZnO or Zn lignosulfonate [147].

Physical treatments	Mode of application	Plant species	Effects	References
Magnetically treated water	Magnetized water (0.32 T), 20 ml of water daily	Chickpea (*Cicer arietinum*)	Increase in plant length, increased photosynthesizing property of plant	[156]
Non-uniform magnetic field	120 mT (rms) for 3 min, 160 mT (rms) for 1 min, and 160 mT (rms) for 5 min.	Lettuce (*Lactuca sativa*)	Improved growth and final yield	[157]
Magnetic seed stimulation	150 mT for 0, 3, 6, 9 and 12 min	Lentil (*Lens culinaris*)	Improved growth of the plant, increased stem length and total mass. Increased root length	[30]
Magnetic seed stimulation	160 mT MF strength for 1 min	Tomato (*Solanum lycopersicum*)	Improved germination	[158]
Stationary magnetic fields	125 and 250 mT for 1, 10 and 20 min, 1 and 24 h and continuous exposure	Pea (*Pisum sativum*)	Stimulating effect on the first stages of growth	[159]
Magnetic seed stimulation	25, 50, 75, 100 and 125 mT for 3 min	Marigold (*Tagetes*)	The results suggest that magnetic field treatments of French marigold seeds have the potential to enhance germination, early growth and biochemical parameters of seedlings	[118]
Magnetic seed stimulation	150 mT for 3 min	Maize (*Zea mays*)	Induced chilling stress tolerance primarily by improving stand establishment, phenology, allometry, agronomic traits and yield components	[60]
Magnetic seed stimulation	25, 50, and 75 mT for 15, 30 and 45 min each	Bitter gourd (*Momordica charantia*)	Improved emergence, growth, yield and yield related parameters	[61]
Low-temperature cold Plasma Treatment	Seeds were coated with tetra fluoride (CF_4) or octa deca fluoro decalin (ODFD)	Radish (*Raphanus sativus*)	60% in the sprout lengths of radish exposed to oxygen RF plasma	[41]
Low-temperature cold plasma treatment	Seeds were coated with tetra fluoride (CF_4) or octadecafluorodecalin (ODFD)	Pea (*Pisum sativum*)	The germination rate was increased by 30% whereas no change was seen in the seeds treated at higher pressure relative to control for the same treatment time	[41]

Physical treatments	Mode of application	Plant species	Effects	References
Gamma rays	0.17 kGy with dose rate 2.18 kGy/h)	Wheat (*Triticum aestivum*)	Increase in percentage emergence	[160]
UV-C	254 nm for 60 min	Ground nut (*Arachis hypogaea*)	Effect on seed germination, seedling growth and productivity	[161]
Gamma rays	10, 15, 20, 25 and 30 Krad doses	*Brassica napus*	An increase was noticed for number of seeds/siliqua, protein and oil contents	[162]

Table 2. Effect of magnetic and radiation treatments on germination, growth and yield of crop plants.

Crop emergence, crop growth and development of two pea varieties with a significant increase in the seed yield have been reported. It was reported that contents of sugar were increased with magnetic seed stimulation in sugar beet roots, and gluten contents were also increased in wheat kernals when magnetic field was applied to the seeds before sowing [148]. Similarly, many researchers had reported higher grain yield due to improved stand establishment, growth and development in agronomic and horticultural crops (**Table 2**).

Nonetheless, priming effects are not only limited to stand establishment and yield, water productivity and uptake of beneficial minerals with reduction in harmful ion but the quality of harvested produce is also improved [60, 131, 135]. Thus, it offers promising and economical solution to improve crop resistance against low and high temperature, flooding and drought, salinity and nutrient stress and effective strategies for agronomic biofortification when combined with soil management and crop genetics.

5. Conclusions and future prospects

Seed enhancements have a wide range of commercial applications from improved crop stands through better germination rates and seedling vigour effective in crop stress management, and improved crop yields together with efficient use of resources such as fertilizers, water and seeds. Sustainable crop production requires the adoption of low-cost and environment friendly seed enhancement techniques. Biological seed enhancement with bacteria and fungi is one of the most appropriate techniques in disease control and growth promotion which can be exploited by seed industry.

The biochemical pathways by which these techniques affect different processes regulating growth and development need to be elucidated.

Longevity of primed seeds during storage remains a problem, which needs to be re-addressed, and work should be extended on other physical or biological seed treatments for their storability.

Nutrient priming with micronutrients not only help to overcome seedling constraints but can also be applied as a complementary approach for biofortification to harvest grains high in Fe, Zn and Mn. Priming invokes stress tolerance and improves performance of varieties containing QTL for stress tolerance such as Swarna containing Sub1 for submergence tolerance and IR74 containing Pup1 for high phosphorus uptake. The integration of molecular approaches with seed enhancement may significantly contribute to seed vigour and results may be delivered to the next generation of seed.

Author details

Irfan Afzal[1*], Hafeez Ur Rehman[1], Muhammad Naveed[2] and Shahzad Maqsood Ahmed Basra[1]

*Address all correspondence to: iafzal@uaf.edu.pk

1 Department of Agronomy, University of Agriculture, Faisalabad, Pakistan

2 Institute of Soil and Environmental Sciences, University of Agriculture, Faisalabad, Pakistan

References

[1] Afzal, I. 2013. Need for public-private partnership in seed technology. Dawn, October 14, 2013.

[2] Burris, J.S. 1980. Maintenance of soybean seed quality in storage as influenced by moisture, temperature and genotype. Iowa State J. Res. 54: 377–389.

[3] Tatipata, A. 2009. Effect of seed moisture content packaging and storage period on microchondria inner membrane of soybean seed. J. Agric. Technol. 5(1): 51–54.

[4] Halmer, P. 2003. Methods to improve seed performance. In: Benech–Arnold, RL, 24 Sanchez, RA (Eds.). Seed Physiology, Applications to Agriculture. Food Product Press, 25 New York, NY.

[5] Ellis, R.H. 2004. Seed and seedling vigour in relation to crop growth and yield. J. Regul. 11(3): 249–255.

[6] Ventura, L., Donà, M., Macovei, A., Carbonera, D., Buttafava, A., Mondoni, A., Rossi, G, Balestrazzi, A. 2012. Understanding the molecular pathways associated with seed vigor. Plant Physiol. Biochem. 60: 196–206.

[7] Doijode, S.D. 2006. Seed quality in vegetable crops. In: Basra, AS (Ed.). Handbook of Seed Science and Technology, The Haworth Press, NY, USA. pp. 677–702.

[8] Gill, T.B., Bates, R., Bicksler, A., Burnette, R., Ricciardi, V., Yoder, L. 2013. Strengthening informal seed systems to enhance food security in Southeast Asia. J. Agric. Food Syst. Commun. Develop. 3(3): 139–153. http://dx.doi.org/10.5304/jafscd.2013.033.005

[9] Koster, K.L., Leopold, A.C. 1988. Sugars and desiccation tolerance in seeds. Plant Physiol. 88: 829–832.

[10] Leprince, O., Harren, F.J.M., Buitink, J., Alberda, M., Hoekstra, F.A. 2000. Metabolic dysfunction and unabated respiration precede the loss of membrane integrity during dehydration of germinating radicles. Plant Physiol. 122: 597–608.

[11] Buitink J., Ly Vu, B., Satour, P., Leprince, O. 2003. A physiological model to study the re-establishment of desiccation tolerance in germinated radicles of Medicago truncatula Gaertn. seeds. Seed Sci. Res. 13: 273–286.

[12] Hadas, A. 2004. Seedbed preparation: the soil physical environment of germinating seeds. In: Benech-Arnold, R.L., Sanchez, R.A. (Eds.). Handbook of Seed Physiology: Applications to Agriculture. Food Product Press, New York, NY. pp. 3–36.

[13] Harris, D. 1996. The effects of manure, genotype, seed priming, depth and date of sowing on the emergence and early growth of *Sorghum bicolor* (L.) Moench in semi-arid Botswana. Soil Tillage Res. 40: 73–88.

[14] Bewley, J.D., Bradford, K.J., Hilhorst, H.W.M., Nonogaki, H. 2013. Seeds: Physiology of Development, Germination and Dormancy, Third Edition. Springer, New York, NY.

[15] Heydecker, W., Coolbear, P. 1977. Seed treatments for improved performance—survey and attempted prognosis. Seed Sci. Technol. 3: 353–425.

[16] Bradford, K.J. 1986. Manipulation of seed water relations via osmotic priming to improve germination under different field conditions. Res. J. Agric. Biol. Sci. 22: 33–37.

[17] Taylor, A.G., Allen, P.S., Bennett, M.A., Bradford, K.J., Burris, J.S., Misra, M.K. 1998. Seed enhancements. Seed Sci. Res. 8: 245–256.

[18] Halmer, P. 2000. Commercial seed treatment technology. In: Black, M., Bewley, J.D. (Eds.). Seed Technology and Its Biological Basis. Sheffield Academic Press, Sheffield, UK. pp. 257–286.

[19] Belal, E.B., Hassan, M.M., El-Ramady, H.R. 2013. Phylogenetic and characterization of salt-tolerant rhizobial strain nodulating faba bean plants. Afr. J. Biotechnol. 12(27): 4324–4337.

[20] Hanci, F., Cebeci, E., Polat, Z. 2014. The effects of Trichoderma harzianum on germination of onion (*Allium cepa* L.) seeds. TABAD Res. J. Agric. Sci. 7 (1): 45–48.

[21] Halmer, P. 2008. Seed technology and seed enhancement. Acta Hortic. 771: 17–26.

[22] Bray, C.M. 1995. Biochemical processes during the osmoconditioning of seeds. In: Kigel, J., Galili, G. (Eds.), Seed Development and Germination. Marcel Dekker, New York, NY. pp. 767–789.

[23] McDonald, M. 2000. Seed priming. In: Black, M., Bewley, J.D. (Eds.). Seed Technology and Its Biological Basis .Sheffield Academic Press, Sheffield, UK. pp. 287–325.

[24] Pietruszewski, S., Martínez, E. 2015. Magnetic field as a method of improving the quality of sowing material: a review. Int. Agrophys. 29: 377–389.

[25] SCOPUS. 2015. https://www.scopus.com

[26] Afzal, I., Basra, S.M.A., Cheema, M.A., Haq, M.A., Kazmi, M.H., Irfan, S. 2011. Enhancement of antioxidant defense system induced by hormonal priming in wheat. Cereal Res. Commun. 39(3): 334–342.

[27] Farooq, M., S.M.A. Basra, R. Tabassum, I. Afzal. 2006. Enhancing the performance of direct seeded fine rice by seed priming. Plant Prod. Sci. 9(4): 446–456.

[28] Araújo, S.S., Paparella, S., Dondi, D., Bentivoglio, A., Carbonera, D., Balestrazzi, A. 2016. Physical methods for seed invigoration: advantages and challenges in seed technology. Front. Plant Sci. 7: 646. doi: 10.3389/fpls.2016.00646

[29] Silva, J.A.T., Dobranszki, J. 2016. Magnetic fields: how is plant growth and development impacted? Protoplasma 253: 231–248.

[30] Aladjadjiyan, A. 2010. Influence of stationary magnetic field on lentil seeds. Int. Agrophys. 24: 321–324.

[31] Pietruszewski, S., Kania, K. 2010. Effect of magnetic field on germination and yield of wheat. Int. Agrophys. 24, 275–302.

[32] Balouchi H.R. and Sanavy S.A.M.M., 2009. Electromagnetic field impact on annual medics and dodder seed germination Int. Agrophysics, 23, 111–115

[33] Yinan, Y., Yuan, L., Yougqing, Y., Chunyang, L. 2005. Effect of seed pretreatment by magnetic field on the sensitivity of cucumber (*Cucumis sativus*) seedlings to ultraviolet-B radiation. Environ. Exp. Bot. 54: 286–294.

[34] De Souza, A., Garcia, D., Sueiro, L., Gilart, F., Porras, E., Licea, L. 2006. Pre-sowing magnetic seed treatments of tomato seeds increase the growth and yield of plants. J. Bioelectric. 27: 247–257.

[35] Aladjadjiyan, A. 2002. Study of the influence of magnetic field on some biological characteristics of maize (*Zea mays* L.). J. Cent. Eur. Agric. 3: 89–94.

[36] Morejon, P., Castro Palacio, J.C., Velazquez Abad, L., Govea, A.P. 2007. Simulation of Pinus Tropicalism seeds by magnetically treated water. Int. Agrophys. 21:173–177.

[37] Sosnin, E.A., Stoffels, E., Erofeev, M.V., Kieft, I.E., Kunts, S.E. 2004. The effects of UV irradiation and gas plasma treatment on living mammalian cells and bacteria: a comparative approach. Plasma Sci., IEEE Trans. 32: 1544–1550.

[38] Akitsu, T., Ohkawa, H., Tsuji, M., Kimura, H., Kogoma, M. 2005. Plasma sterilization using glow discharge at atmospheric pressure. Surface Coatings Technol. 193: 29–34.

[39] Hayashi, N., Nakahigashi, A., Goto, M., Kitazaki, S., Koga, K., Shiratani, M. 2011. Redox characteristics of thiol compounds using radicals produced by water vapor radio frequency discharge. Japanese J. Appl. Phys. 50: 08JF04.

[40] Klämpfl, T.G., Isbary, G., Shimizu, T., Li, Y.-F., Zimmermann, J.L., Stolz, W., Schlegel, J., Morfill, G.E., Schmidt, H.-U. 2012. Cold atmospheric air plasma sterilization against spores and other microorganisms of clinical interest. Appl. Environ. Microbiol. 78: 5077–5082.

[41] Volin, J.C., Denes, F.S., Young, R.A., Park, S.M. 2000. Modification of seed germination performance through cold plasma chemistry technology. Crop Sci. 40: 1706–1718.

[42] Jiayun, T., Rui, H., Xiaoli, Z., Ruoting, Z., Weiwen, C., Size, Y. 2014. Effects of atmospheric pressure air plasma pretreatment on the seed germination and early growth of Andrographis paniculata. Plasma Sci. Technol. 16: 260.

[43] Šerá, B., Špatenka, P., Šerý, M., Vrchotova, N., Hrušková, I. 2010. Influence of plasma treatment on wheat and oat germination and early growth. Plasma Sci., IEEE Trans. 38: 2963–2968.

[44] Zhou, Z., Huang, Y., Yang, S., Chen, W. 2011. Introduction of a new atmospheric pressure plasma device and application on tomato seeds. Agric. Sci. 2: 23.

[45] Henselová, M., Slováková, Ľ., Martinka, M., Zahoranová, A. 2012. Growth, anatomy and enzyme activity changes in maize roots induced by treatment of seeds with low-temperature plasma. Biologia. 67: 490–497.

[46] Dhayal, M., Lee, S.-Y., Park, S.-U. 2006. Using low-pressure plasma for Carthamus tinctorium L. seed surface modification. Vacuum. 80: 499–506.

[47] Selcuk, M., Oksuz, L., Basaran, P. 2008. Decontamination of grains and legumes infected with Aspergillus spp. and Penicillum spp. by cold plasma treatment. Bioresour. Technol. 99, 5104–5109.

[48] Lynikiene, S., Pozeliene, A., Rutkauskas, G. 2006. Influence of corona discharge field on seed viability and dynamics of germination. Int. Agrophys. 20: 195.

[49] Bormashenko, E., Grynyov, R., Bormashenko, Y., Drori, E. 2012. Cold radiofrequency 9 plasma treatment modifies wettability and germination speed of plant seeds. Sci. Rep. 10 2: 741–748.

[50] Filatova, I., Azharonok, V., Lushkevich, V., Zhukovsky, A., Gadzhieva, G., Spasić, K., Živković, S., Puač, N., Lazović, S., Malović, G. 2013. Plasma seeds treatment as a

promising technique for seed germination improvement. 31st International Conference on Phenomena in Ionized Gases, Granada, Spain.

[51] Ling, L., Jiafeng, J., Jiangang, L., Minchong, S., Xin, H., Hanliang, S. et al. 2014. Effects of cold plasma treatment on seed germination and seedling growth of soybean. Sci. Rep. 4: 5859. http://dx.doi.org/10.1038/srep05859

[52] Kiong, A., Lai, A., Hussein, S., Harun, A.R. 2008. Physiological responses of *Orthosiphon stamineus* plantlets to gamma irradiation. Am-Eurasian. J. Sustain. Agric. 2: 135–149.

[53] Melki, M., Salami, D. 2008. Studies the effects of low dose of gamma rays on the behavior of *Chickpea* under various conditions. Pak. J. Biol. Sci. 11(19): 2326–2330.

[54] Rogozhin, V.V., Kuriliuk, T.T., Filippova, N.P. 2000. Change in the reaction of the antioxidant system of wheat sprouts after UV-irradiation of seeds. Biofizika. 45: 730–736.

[55] Wi, S.G., Chung, B.Y., Kim, J.S. et al. 2007. Effects of gamma irradiation on morphological changes and biological responses in plants. Micron. 38: 553–564.

[56] Ashraf, M. 2009. Biotechnological approach of improving plant salt tolerance using antioxidants as markers. Biotechnol. Adv. 27: 84–93.

[57] Bhosale, R.S., More, A.D. 2013b. Effect of EMS (ethyl methane sulfonate) on seed germination, seedling height and seedling injury in *Withania somnifera*, (L.) Dunal., Int. J. Life Sci. 1(2): 158–160.

[58] Yilmaz, A., Boydak, E. 2006. The effects of cobalt-60 applications on yield components of cotton (*Gossypium barbadense* L.). Pak. J. Biol. Sci. 9(15): 2761–2769.

[59] Afzal, I., Ashraf, S., Qasim, M., Basra, S.M.A. Shahid, M. 2009. Does halopriming improve germination and seedling vigor in marigold (Tagetus sp.). Seed Sci. Technol. 37: 436–445.

[60] Afzal, I., Noor, M.A., Bakhtavar, M.A., Ahmad, A. Haq, Z. 2015. Improvement of spring maize (*Zea mays*) performance through physical and physiological seed enhancements. Seed Sci. Technol. 43(2): 1–12.

[61] Iqbal, M., Haq, Z., Malik, A., Ayub, C.M., Jamil, Y. Nisar, J. 2016. Pre-sowing seed magnetic field stimulation: a good option to enhance bitter gourd germination, seedling growth and yield characteristics. Biocatal. Agric. Biotechnol. 5: 30–37.

[62] Bakhtavar, M.A., Afzal, I., Basra, S.M.A., Ahmad, A., Noor, M. 2015. Physiological strategies to improve the performance of spring maize (*Zea mays* L.) planted under early and optimum sowing conditions. PLoS ONE. 10(4): e0124441.

[63] Khan, A.A. 1992. Preplant physiological seed conditioning. Horticult. Rev. 13: 131–181.

[64] Pill, W.G., Frett, J.J., Morneau, D.C. 1991. Germination and seedling emergence of primed tomato and asparagus seeds under adverse conditions. HortScience. 26: 1160–1162.

[65] Afzal, I., Hussain, B., Basra, S.M.A., Rehman, H. 2012. Priming with MLE reduces imbibitional chilling injury in spring maize. Seed Sci. Technol. 40: 271–276.

[66] Pill, W.G., Necker, A.D. 2001. The effect of seed treatment on germination and establishment of Kentucky blue grass (*Poa pretenses* L.). Seed Sci. Technol. 29: 65–72.

[67] Janmohammadi, M., Moradi Dezfuli, P., Sharifzadeh, F. 2008. Seed invigoration techniques to improve germination and early growth of inbred line of maize under salinity and drought stress. Gen. Appl. Plant Physiol. 34(3–4): 215–226.

[68] Farooq M., S.M.A. Basra, I. Afzal and A. Khaliq. 2006. Optimization of hydropriming techniques for rice seed invigoration. Seed Sci. Technol., 34: 507–512.

[69] Karssen, C.M., Zagorski, S., Kepczynski, J., Groot, S.P.C. 1989. Key role for endogenous gibberellins in the control of seed germination. Ann. Bot. 63:71–80.

[70] Riefler, M., Novak, O., Strnad, M., Schmülling, T. 2006. *Arabidopsis* cytokinin receptor mutants reveal functions in shoot growth, leaf senescence, seed size, germination, root development, and cytokinin metabolism. Plant Cell. 18: 40–54.

[71] Rood, S.B., Williams, P.H., Pearce, D., Murofushi, N., Mander, L.N., Pharis, R. 1990. A mutant gene that increases gibberellin production in *Brassica*. Plant Physiol. 93: 1168–1174.

[72] Foidle, N., Makkar, H.P.S., Becker, K. 2001. The potential of *Moringa oleifera* for agricultural and industrial uses. In: Fuglie, L. (Ed.). The Miracle Tree: The Multipurpose Attributes of Moringa. CTA Publications, Wageningen, The Netherlands. pp. 45–76.

[73] Farooq, M., Wahid, A. Siddique, K.H.M. 2012. Micronutrient application through seed treatments—a review. J. Soil Sci. Plant Nutrition. 12(1): 125–142.

[74] Wilhelm, N.S., Graham, R.D., Rovira, A.D. 1988. Application of different sources of manganese sulphate decreases take-all (*Gaeumannomyces graminis* var. *tritici*) of wheat grown in a manganese deficient soil. Austr. J. Agric. Res. 39: 1–10.

[75] Peeran, S.N., Natanasabapathy, S. 1980. Potassium chloride pretreatment on rice seeds. Int. Rice Res. Newsletter. 5: 19.

[76] Sherrell, C.G. 1984. Effect of molybdenum concentration in the seed on the response of pasture legumes to molybdenum. New Zealand J. Agric. Res. 27: 417–423.

[77] Harris D, Raghuwanshi BS, Gangwar JS, Singh SC, Joshi KD, Rashid A and Hollington PA (2001). Participatory evaluation by farmers of 'on-farm' seed priming in wheat in India, Nepal and Pakistan. Experimental Agriculture 37 (3): 403-415.

[78] Glick, B.R., Penrose, D.M., Li, J. 1998. A model for the lowering of plant ethylene concentrations by plant growth promoting bacteria. J. Theor. Biol. 190: 63–68.

[79] Nadeem, S.M., Naveed, M., Ahmad, M. Zahir, Z.A. 2015. Rhizosphere bacteria for biomass production and improvement of stress tolerance: mechanisms of action, applications and future prospects. In: Arora, N.K. (Ed.). Plant Microbe Symbiosis—Applied Facets. Springer, The Netherlands. pp. 1–36.

[80] Bloemberg, G.V., Lugtenberg, B.J. 2001. Molecular basis of plant growth promotion and biocontrol by rhizobacteria. Curr. Opin. Plant Biol. 4(4): 343–350.

[81] Vessey, J.K. 2003. Plant growth promoting rhizobacteria as biofertilizers. Plant Soil. 255(2): 571–586.

[82] Burns, T.A., Bishop, P.E., Israel, D.W. 1981. Enhanced nodulation of leguminous plant roots by mixed cultures of Azotobacter vinelandii and Rhizobium. Plant Soil. 62: 399–412.

[83] Yahalom, E., Okon, Y., Dovrat, A. 1987. *Azospirillum* effects on susceptibility to *Rhizobium* nodulation and on nitrogen fixation of several forage legumes. Can. J. Microbiol. 33: 510–514.

[84] Junges, E., Muniz, M.F.B., Bastos, B.d.O., Oruoski, P. 2016. Biopriming in bean seeds, Acta Agric. Scand., Sect. B: Soil Plant Sci. 66: 207–214.

[85] Nadeem, S.M., Zahir, Z.A., Naveed, M., Arshad, M. 2007. Preliminary investigations on inducing salt tolerance in maize through inoculation with rhizobacteria containing ACC deaminase activity. Can. J. Microbiol. 53: 1141–1149.

[86] Zahir, Z.A., Zafar-ul-Hye, M., Sajjad, S. Naveed, M. 2011. Comparative effectiveness of *Pseudomonas* and *Serratia* sp. containing ACC-deaminase for coinoculation with *Rhizobium leguminosarum* to improve growth, nodulation, and yield of lentil. Biol. Fertility Soils. 47: 457–465.

[87] Li, DM, Alexander, M. 1988. Co-inoculation with antibiotic producing bacteria to increase colonization and nodulation by rhizobia. Plant Soil. 108: 211–219.

[88] Grover, M., Ali, Sk.Z., Sandhya, V., Rasul, A., Venkateswarlu, B. 2011. Role of micro-organisms in adaptation of agriculture crops to abiotic stresses. World J. Microbiol. Biotechnol. 27: 1231–1240.

[89] Glick, B.R. Plant Growth-Promoting Bacteria: Mechanisms and Applications; Hindawi Publishing Corporation, Scientifica: Waterloo, Canada, 2012

[90] Mitter, B., Brader, G., Afzal, M., Compant, S., Naveed, M., Trognitz, F. Sessitsch, A. 2013. Advances in elucidating beneficial interactions between plants, soil and bacteria. Adv. Agron. 121: 381–445.

[91] Balbinot, E., Lopes, H.M. 2006. Effects of priming and drying on germination and vigor of carrot seeds. Revista Brasileira de Sementes, Pelotas. 13 28: 1–8.

[92] Cayuela, E., Pérez-Alfocea, F., Caro, M., Bolarin, M.C. 1996. Priming of seeds with NaCl induces physiological changes in tomato plants grown under salt stress. Physiol. Plantarum. 96(2): 231–236.

[93] Amjad, M., Ziaf, K., Iqbal, Q., Ahmad, I., Riaz, M.A., Saqib, Z.A. 2007. Effect of seed priming on seed vigour and salt tolerance in hot pepper. Pak. J. Agri. Sci. 44(3): 408–416.

[94] Mariselvam, D. 2012. Performance of bioprimed bhendi (cv. arka anamika) seeds with biocontrol agents and liquid biofertilizers under laboratory and field conditions. M. Sc. (Ag.) Thesis, Tamil Nadu Agricultural University, Coimbatore.

[95] Hedden, P., Phillips, A.L. 2000. Gibberellin metabolism: new insights revealed by the genes. Trends Plant Sci. 5(12): 523–530.

[96] Arkhipova, T.N., Veselov, S.U., Melentiev, A.I., Martynenko, E.V., Kudoyarova, G.R. 2005. Ability of bacterium *Bacillus subtilis* to produce cytokinins and to influence the growth and endogenous hormone content of lettuce plants. Plant Soil. 272: 201–209.

[97] Van Loon, L.C., Bakker, P.A.H.M., Pieterse, C.M.J. 1998. Systemic resistance induced by rhizosphere bacteria. Annu. Rev. Phytopathol. 36: 453–483.

[98] Weller, D.M. 1988. Biological control of soilborne plant pathogens in the rhizosphere with bacteria. Ann. Rev. Phytopathol. 26(1): 379–407.

[99] Begum, M.F., Rahman, M.A., Alam, M.F. 2010. Biological control of Alternaria fruit rot of chili by *Trichoderma* species under field conditions. Mycobiology. 38(2): 113–117.

[100] El-Mohamedy, R.S.R., Abd Alla, M.A. 2013. Bio-priming seed treatment for biological control of soil borne fungi causing root rot of green bean (*Phaseolus vulgaris* L.). J. Agric. Technol. 9: 589–599.

[101] Marulanda, A., Porcel, R., Barea, J., Azcón, R., 2007. Drought tolerance and antioxidant activities in lavender plants colonized by native drought-tolerant or drought-sensitive Glomus species. Microb. Ecol. 54: 543–552.

[102] Kasim, W.A., Osman, M.E., Omar, M.N., Abd El-Daim, I.A., Bejai, S., Meijer, J., 2013. Control of drought stress in wheat using plant growth-promoting bacteria. J. Plant Growth Regul. 32: 122–130.

[103] Bhatt, R.M., Selvakumar, G., Upreti, K.K., Boregowda, P.C. 2015. Effect of biopriming with *Enterobacter* strains on seed germination and seedling growth of tomato (*Solanum lycopersicum* L.) under osmotic stress. Proc. Natl. Acad. Sci., India, Sect. B Biol. Sci. 85(1): 63–69.

[104] Thomas, B., Murphy, D.J., Murray, B.G. 2003. Encyclopedia of Applied Plant Sciences (3 volume set). Elsevier Science BV.

[105] Stendahl, F., 2005. Seed Coating for Delayed Germination. A tool for relay cropping of annual crops. Licentiate thesis. Department of Ecology and Crop Production Science, Swedish University of Agricultural Sciences, Uppsala, Sweden.

[106] Scott, J.M. 1989. Seed Coatings and Treatments and their Effects on Plant Establishment. Advances in Agronomy 42: 43–83.

[107] Rehman, A., Farooq, M., Ata, Z., Wahid, A. 2013. Role of Boron in leaf elongation and tillering dynamics in fine-grain aromatic rice. J. Plant Nutr. 36(1): 42–54.

[108] Ni, B. R. 1997. Seed coating, film coating and pelleting. In: Seed Industry and Agricultural Development, Chinese Association of Agricultural Sciences, DOA, Ministry of Agriculture, Beijing, China Agriculture Press, pp. 737–747.

[109] Sikhaoa, P., Taylor, A.G., Marinoa, E.T., Catranisa, C.M., Sirib, B. 2015. Development of seed agglomeration technology using lettuce and tomato as model vegetable crop seeds. Sci. Horticulturae. 184: 85–92.

[110] Hare, P.D., Cress, W.A., Van Staden, J. 1998. Dissecting the roles of osmolyte accumulation during stress. Plant Cell Environ. 21: 535–553.

[111] Hasegawa, P.M., Bressan, R.A., Zhu, J.K., Bohnert, H.J. 2000. Plant cellular and molecular responses to high salinity. Annu. Rev. Plant Physiol. Plant Mol. Biol. 51: 463–499.

[112] Iqbal, M., Ashraf, M. 2010. Changes in hormonal balance: a possible mechanism of pre-sowing chilling-induced salt tolerance in spring wheat. J. Agr. Crop Sci. 196: 440–454.

[113] Ella ES, Dionisio–Sese ML, Ismail AM. 2011. Seed pre-treatment in rice reduces damage, enhances carbohydrate mobilization and improves emergence and seedling establishment under flooded conditions. AoB Plants plr007, doi:10.1093/aobpla/plr007

[114] Gorim, L., Asch, F. 2015. Seed coating reduces respiration losses and affects sugar metabolism during germination and early seedling growth in cereals. Funct. Plant Biol. 42: 209–218.

[115] Rochalska, M., Grabowska, K. 2007. Influence of magnetic fields on activity of enzyme: a- and b-amylase and gluta-thione S-transferase (GST) in wheat plants. Int. Agrophys. 21: 185–188.

[116] Mamat, H., Aini, I.N., Said, M., Jamaludin, R. 2005. Physico-chemical characteristics of palm oil and sunflower oil blends fractionated at different temperatures. Food Chem. 91: 731–736.

[117] Zia ul Haq, Z., Jamil, Y., Irum, S., Randhawa, M.A., Iqbal, M. Amin, N. 2012. Enhancement in the germination, seedling growth and yield of radish (*Raphanus sativus*) using seed pre-sowing magnetic field treatment. Polish J. Environ. Stud. 21: 369–374.

[118] Afzal, I., Mukhtar, K., Qasim, M., Basra, S.M.A., Shahid, M. Haq, Z. 2012. Magnetic stimulation of marigold seed. Int. Agrophysics. 26(3). 335–339.

[119] Asgedom, H., Becker, M. 2001. Effect of seed priming with nutrient solutions on germination, seedling growth and weed competitiveness of cereals in Eritrea. In: Proc. Deutscher Tropentag 2001, Univ. Bonn and ATSAF, Margraf Publishers Press, Weickersheim. p. 282.

[120] de Castro, R.D., van Lammeren, A.A.M., Groot, S.P.C., Bino, R.J., Hilhorst, H.W.M. 2000. Cell division and subsequent radicle protrusion in tomato seeds are inhibited by osmotic stress but DNA synthesis and formation of microtubular cytoskeleton are not. Plant Physiol. 122: 327–335.

[121] Coolbear, P. et al. 1980, "Osmotic pre-sowing treatments and nucleic acid accumulation in tomato seeds", (Lycopersicon Cycopersicum), Seed Sci. Technol. 8, 289–303.

[122] Burgass, R.W., Powell, A.A. 1984. Evidence for repair processes in the invigoration of seeds by hydration. Ann. Botany. 53, 753–757.

[123] Gallardo, K., Job, C., Groot, S.P.C., Puype, M., Demol, H., Vandekerckhove, J., Job, D. 2001. Proteomic analysis of Arabidopsis seed germination and priming. Plant Physiol. 126: 835–848.

[124] Gong, F., Wu, X., Wang, W. 2013. Comparative proteomic identification of embryo proteins associated with hydropriming induced rapid-germination of maize seeds. POJ. 6(5): 333–339.

[125] Schwember, A.R., Bradford, K.J. 2011. Oxygen interacts with priming, moisture content and temperature to affect the longevity of lettuce and onion seeds. Seed Sci. Res. 21: 175–185.

[126] Patade, V.Y., Khatri, D., Manoj, K., Kumari, M., Ahmed, Z. 2012. Cold tolerance in thiourea primed capsicum seedlings is associated with transcript regulation of stress responsive genes. Mol. Biol. Rep. 39: 10603–10613.

[127] Imran, M., Mahmood, A., Römheld, V., Neumann, G. 2013. Nutrient seed priming improves seedling development of maize exposed to low root zone temperatures during early growth. Eur. J. Agron. 49: 141–148.

[128] Nautiyal, N., Shukla, K. Evaluation of seed priming zinc treatments in chickpea for seedling establishment under zinc deficient conditions. J. Plant Nutr. 36: 251–258.

[129] Marcar, N.E., Graham, R.D. 1986. Effect of seed manganese content on the growth of wheat (*Triticum aestivum*) under manganese deficiency. Plant Soil. 96: 165–173.

[130] Raj, A.S. 1987. Cobalt nutrition of pigeonpea and peanut in relation to growth and yield. J. Plant Nutr. 10: 2137–2145.

[131] Rehman, H., Iqbal, H., Basra, S.M.A., Afzal, I., Farooq, M., Wakeel, A., Wang, N. 2015. Seed priming improves early vigor improved growth and productivity of spring maize. J. Integr. Agric. 14(9): 1745–1754.

[132] Anwar, M.P., Juraimi, A.S., Puteh, A., Selamat,A., Rahman, M.M, Samedani, B. 2012. Seed priming influences weed competitiveness and productivity of aerobic rice. Acta Agric. Scand. 62: 499–509.

[133] Pame, A.R., Kreye, C., Johnson, D., Heuer, S. Becker, M. 2015. Effects of genotype, seed P concentration and seed priming on seedling vigor of rice. Expl. Agric. 51(3): 370–381.

[134] Hasni, Z.A., Chaudhary, M.A., Niaz, J. 1985. Effect of magnetic field seed treatment on wheat. J. Pak. Agri. Sci. 22: 33–37.

[135] Rehman, H., Kamran, M., Basra, S.M.A., Afzal, I., Farooq, M. 2015. Influence of seed priming on the performance and water productivity of direct seeded rice in alternating wetting and drying. Rice Sci. 22(4): 189–196.

[136] Johnson, S.E., Lauren, J.G., Welch, R.M., Duxbury, J.M. 2005. A comparison of the effects of micronutrient seed priming and soil fertilization on the mineral nutrition of chickpea (*Cicer arietinum*), lentil (*Lens culinaris*), rice (*Oryza sativa*) and wheat (*Triticum aestivum*) in Nepal. Exp. Agric. 41: 427–448.

[137] Subedi, K.D., Ma, B.L. 2005. Seed priming does not improve corn yield in a humid temperate environment. Agron. J. 97: 211–218.

[138] Sarkar, R.K. 2012. Seed priming improves agronomic trait performance under flooding and non-flooding conditions in rice with QTL SUB1. Rice Sci. 19: 286–294.

[139] Mahajan, G., Sarlach, R.S., Japinder, S., Gill, M.S. 2011. Seed priming effects on germination, growth and yield of dry direct-seeded rice. J. Crop Improv. 25(4): 409–417.

[140] Slaton, N.A., Wilson, C.E. JR, Ntamatungiro, S., Norman, R.J., Boothe, D.L. 2001. Evaluation of zinc seed treatments for rice. Agron. J. 93: 157–163.

[141] Khalid, F., Ahmad, A.-U.-H., Farooq, M., Murtaza, G. 2015. Evaluating the role of seed priming in improving the performance of nursery seedlings for system of rice intensification. Pak. J. Agric. Sci. 52: 27–36.

[142] Nawaz, A., Farooq, M., Ahmad, R., Maqsood, S. Basra, A., Lal, A. 2016. Seed priming improves stand establishment and productivity of no. till wheat grown after direct seeded aerobic and transplanted flooded rice. Eur. J. Agron. 76: 130–137.

[143] Rehman, A., Farooq, M., Nawaz, A., Ahmad, R. 2016. Improving the performance of short-duration basmati rice in water-saving production systems by boron nutrition. Ann. Appl. Biol. 168(1): 19–28.

[144] Iqbal, M., Ashraf, M. 2013. Alleviation of salinity-induced perturbations in ionic and hormonal concentrations in spring wheat through seed preconditioning in synthetic auxins. Acta Physiol. Plant. 35: 1093–1112.

[145] Shah, Z., Haq, I.u., Rehman, A., Khan, A., Afzal, M. 2013. Soil amendments and seed priming influence nutrients uptake, soil properties, yield and yield components of wheat (*Triticum aestivum* L.) in alkali soils. Soil Sci. Plant Nutr. 59: 262–270.

[146] Hussian, I., Ahmad, R., Farooq, M. Wahid, A. 2013. Seed priming improves the performance of poor quality wheat seed. Int. J. Agric. Biol. 15: 1343–1348.

[147] Mengel, D.B., Wilson, F.E. 1979. Correction of Zn deficiency in direct seeded rice. Int. Rice Res. Newsl. 4: 24–25.

[148] Pietruszewski, S., Wójcik, S. 2000. Effect of magnetic field on field and chemical composition of sugar beet. Int. Agrophys. 14: 89–92.

[149] Egamberdieva, D. 2010. Growth response of wheat cultivars to bacterial inoculation in calcareous soil. Plant Soil Environ. 56(12): 570–573.

[150] Paul. D., Nair, S. 2008. Stress adaptations in a plant growth rhizobacterium (PGPR) with increasing salinity in the coastal agricultural soils. J. Basic Microbiol. 48: 378–384.

[151] Sandhya, V., Grover, M., Reddy, G., Venkateswarlu, B. 2009. Alleviation of drought stress effects in sunflower seedlings by the exopolysaccharides producing *Pseudomonas putida* strain GAP-P45. Biol. Fertility Soils. 46(1): 17–26.

[152] Bano, A., Fatima, M. 2009. Salt tolerance in *Zea mays* (L). following inoculation with Rhizobium and Pseudomonas. Biol. Fertility Soils. 45(4): 405–413.

[153] Egamberdieva, D. Pseudomonas chlororaphis: a salt–tolerant bacterial inoculant for plant growth stimulation under saline soil conditions. Acta Physiol Plant (2012) 34: 751. doi:10.1007/s11738–011–0875–9

[154] Sun, C., Johnson, J.M., Cai, D., Sherameti, I., Oelmüller, R., Lou, B. 2010. *Piriformospora indica* confers drought tolerance in Chinese cabbage leaves by stimulating antioxidant enzymes, the expression of drought-related genes and the plastid-localized CAS protein. J. Plant Physiol Vol. 167, Pages 1009–1017.

[155] Morse, L.J., Day, T.A., Faeth, S.H. 2002. Effect of Neotyphodium endophyte infection on growth and leaf gas exchange of Arizona fescue under contrasting water availability regimes. Environ. Exp. Bot. 48(3): 257–268.

[156] Nasher, S.H. 2008. The effect of magnetic water on growth of Chick-Pea seeds. Eng. 5 Tech. 26(9).1–4.

[157] De Souza, A., Sueiro, L., Gonzales, L.M., Licea, L., Porras, E.P., Gilart, F. 2008. Improvement of the growth and yield of lettuce plants by non-uniform magnetic fields. Electromagn. Biol. Med. 27: 173–184.

[158] De Souza, Sueiro, A.L., García, D., Porras, E. 2011. Extremely low frequency non-uniform magnetic fields improve tomato seed germination and early seedling growth. Seed Sci. Technol. 38: 61–72.

[159] Carbonell, M.V., Flórez, M., Martínez, E., Maqueda, R., Amaya, J.M. 2011. Study of stationary magnetic fields on initial growth of pea (*Pisum sativum* L.) seeds. Seed Sci. Technol. 39: 673–679.

[160] Groover, S., Khan, A.S. 2014. Effect of ionizing radiation on some characteristics of seeds of wheat. Int. J. Sci. Technol. Res. 3(4): 32–39.

[161] Neelamegam, R. Sutha, T. 2015. UV-C irradiation effect on seed germination, seedling growth and productivity of groundnut (*Arachis hypogaea* L.). Int. J. Curr. Microbiol. Appl. Sci. 4: 430–443.

[162] Khan, W.M., Shah, S.Z., Khan, M.S., Islam, Z.U., Ali, S., Hussain, F., Irshad, M. Zahid, M. 2014. Effects of gamma radiations on some morphological and biochemical characteristics of *Brassica napus* L. (variety Altex). Int. J. Biosci. 4: 36–41.

Chapter 3

Signaling Patterns of Reactive Oxygen Species and Phytohormones During Transition Period of Quiescent Seeds into Metabolically Active Organisms

Prabhakaran Soundararajan,
Abinaya Manivannan and Byoung Ryong Jeong

Additional information is available at the end of the chapter

http://dx.doi.org/10.5772/64789

Abstract

Dormancy and germination of seeds are determined by various factors such as vitality, genotype, hardness, and other environmental cues, such as moisture, air, temperature, and light. Metabolic activity of seeds varies between the quiescent and imbibition state. In the dry state, longevity of a seed is determined by the reactive oxygen species (ROS) such as lipid peroxyl radical (LOO•) and lipid hydroperoxide (LOOH) that are generated nonenzymatically due to lipid peroxidation (LPO). During rehydration phase, enormous amount of ROS, such as superoxide (O_2^-), hydrogen peroxide (H_2O_2) and hydroxyl (•OH) radicals, are generated from the metabolically active compartments such as mitochondria, chloroplasts, and peroxisomes. The progressive conditional, temporal, and spatial distribution of ROS is tightly controlled by the effective antioxidant system that leads to the successful germination of seeds and this phenomenon is defined as 'oxidation window.' Gibberellins (GAs) and abscisic acid (ABA) are the key phytohormones involved in the germination/dormancy. Former promotes germination, whereas the latter induces dormancy. Genes involved in the synthesis and signaling of GA, such as *gibberellin 3-β-dioxygenase* (*GA3ox*), *GA20ox*, and *GA-insensitive dwarf* (*GID*), are responsible for the conversion of GA from an inactive to a bioactive form. On the other hand, DELLA, an important protein family acting as the repressors for GA-regulating genes, is activated by ABA. Function of genes, such as *SLEEPY*, *PICKLE*, SPINDLY, *SECRET AGENT*, *AMYLASE*, *GAMYB*, and *LEAFY*, are interrelated with the GA/ABA metabolism. By inducing the ubiquitin-26S proteolysis pathway, GA overcomes the DELLA-mediated effects on germination. The E3 ubiquitin ligase SCF^{SLY1} (skp1-cullin-F-box-$Rbx1^{SLY1}$) complex was reported to be involved in the degradation of DELLA proteins. Additionally, cell differentiation and elongation process sustained by the ROS were also linked with the ethylene, brassinosteroids, and auxins. Hence, this chapter provides the heuristic framework on the phenomenon of

INTECH
open science | open minds

systemic cross-talk between the ROS and phytohormones during the transition period of quiescent seeds into the metabolically active organisms.

Keywords: dormancy, germination, metabolism, oxygen radicals, signaling, mechanism

1. Introduction

After pollination (double fertilization), the typical diploid embryos are covered by the triploid endosperm and the diploid testa. The triploid endosperm consists of nutritive tissues and living cells, while the diploid testa includes seed coat, maternal tissue, and dead cells. Seeds are the vital component allowing embryo dispersion and its consequent development into mature plants [1]. The seeds of monocots and dicots differ in their structure and method of emergence [1, 2]. However, here we comprehensively focus on the signaling pattern of the reactive oxygen species (ROS) and its interaction during the germination and dormancy condition.

Although the dispersal of seeds is absolutely dependent on various cues such as vitality, genotype, hardness, moisture, air, temperature, light, and duration of seed storage, endosperm weakening is one of the key factors that determine the protrusion of radicle. The term "coat-associated dormancy" refers to the mechanical constraint that can impair germination, while "embryo dormancy" is characterized by the embryo failure to develop [1–3].

After imbibition, the weakening of endosperm is dependent on the gas exchange/respiration. Loosening of the endosperm was suggested to be influenced by the proper localization of ROS and its fine regulation by the antioxidant systems [4, 5]. The proper reduction/oxidation of the ROS (redox homeostasis) plays a key role in the transition from quiescence to active state [6]. Gibberellins (GAs) are involved in the promotion of the endosperm weakening and, on the other hand, abscisic acid (ABA) at least partly inhibits this process either directly or indirectly [2, 3, 7]. Induction/inhibition of the genes responsible for the endosperm weakening is controlled by the ROS-GA-ABA [7–11]. Substantially, cross-talk of other phytohormones, such as ethylene, brassinosteroids and auxins, with GA and ABA were reported to be inevitable for the seed development [12].

Although research in seed biology has reached significant advancements, apparent continuum still lies in the mechanisms underlying germination and dormancy, which needs to be disclosed. In this chapter, the heuristic network on cross-talk between the ROS and phytohormones involved in the release and/or induction of dormancy has been discussed.

2. Seed Respiration

To fulfill the higher energy requirement during the transition period (from quiescent to active state) of the seed, cellular respiration is rapid, high, and synchronized with the mitochondrial activity [13]. According to Law et al., proteins required for the biogenesis of mitochondria were already present in the dried seeds and this process is activated upon imbibition. Although the import of mitochondrial proteins is highly required for the biogenesis, amount of ATP consumption by mitochondria is limited as compared with other processes. On the other hand, amount of ATP required by the mitochondria for the protein import is lesser than the energy required for protein synthesis [14]. In the transition period, oxygen (O_2) released by the respiration governs the internal communications between the cell organelles and the rapid cell division and expansion [15]. The excessive generation of ROS is extremely harmful to the cells. Although the O_2^- has very limited half-life (2 μs), the reduction in O_2^- (superoxide dismutation) results in the production of hydrogen peroxide (H_2O_2). Hydrogen peroxide can travel long distance and reaches the target as its half-life was determined to be about 1 ms. Other free radicals formed during the enzymatic reduction of O_2^- and H_2O_2 is •OH [16]. The formation of •OH radical is mediated by iron in the Haber-Weiss and Fenton reactions. The uncoupled electrons present in the ROS cross-react with other essential metabolites or cellular components, affecting the normal cell physiology. However, a proper antioxidant system detoxifies the excessively generated free radicals and leads to the nondormant phenotype [16]. Detailed mechanisms on the involvement of ROS in seed germination are discussed below.

3. Oxidation Window

During the embryogenesis and the seed-filling process, seeds possessed maximum water content [17]. Subsequently, dramatic water loss takes place in the postmaturation stage [2]. According to the recent reports, ROS do not play a detrimental role during development under controlled conditions [18]. The ROS-mediated signaling is majorly involved in the endosperm weakening, mobilization of seed reserves, programmed cell death (PCD), and also protection against the pathogens [4]. Hence, it can be ascertained that the ROS cannot be simply considered as a hazardous material. Controlled production of ROS and ROS-related molecular interactions represent key factors in various central components of plant biology [16]. In the nondormant seeds, O_2^- and H_2O_2 radicals were uniformly distributed within the radicle, while in dormant seeds irregular patterns were observed. Only seeds with a proper redox homeostasis display nondormant phenotype [7, 11]. Success of the seed germination is apparently associated with the equilibrium between the ROS and its scavenging antioxidant system [1, 4–6]. Uncontrolled generation of ROS is extremely harmful and can lead to several lethal effects. Meanwhile, the tight control over the ROS helps in various developmental processes including germination. This process is generally termed as 'oxidation window' [19].

3.1. Quiescent seed

Quiescent seeds, characterized by low moisture content (5–15%), do not possess active metabolism. During the late embryogenic state, seeds are actively involved in the storage of reserves, while enzymatic activities are gradually decreased. However, during the storage, lipid peroxidation (LPO) occurring on the polyunsaturated fatty acids (PUFA) in the cell also known as respiration burst oxidase homolog (rboh) plays an important role in the transfer of electrons from cytosolic NADPH or NADH to apoplastic oxygen and posttranslational modifications of proteins. In Arabidopsis, *AtrbohD* mutant showed reduced superoxide production and protein carbonylation in dry seeds. However, after imbibition the protein, oxidation level of *AtrbohD* mutant was slightly higher than wild [38]. The posttranslational modifications, especially mRNA oxidations, are governed by the ABA [32].

3.5. Antioxidant enzymes

As mentioned earlier, improper desiccation as well as storage increases the LPO and affects seed vigor. Increased production of ROS from the metabolically active sites during the transition from a quiescent to imbibition state could possibly cause stress. The deleterious effects of the ROS can be overcome by the proper antioxidant system. Both enzymatic and nonenzymatic antioxidants play a vital role in the maintenance of level of the ROS. Rather than complete alleviation, proper activation of antioxidant enzymes directs the ROS to the signaling process [24–27]. Muller et al. reported that ROS are important components in the endosperm weakening [26]. Exogenous application of H_2O_2 or menadione (to generate superoxide) to 3-day old maize seedlings enhances tolerance against the chilling stress [39]. Meanwhile, Pulido et al. found that nuclear localization of peroxiredoxin and thioredoxin prevents nucleic acids from oxidative damage occurring during the maturation and germination in wheat seeds [40]. The detoxification of H_2O_2 in the seed filling is catalyzed by the isoforms of catalase (CAT). The isoform CAT3 is involved highly in the early postpollination, whereas CAT1 and CAT2 isoforms play a crucial role during the seed development [41, 42]. Recently, Leymarie et al. clearly demonstrated the necessity of the ROS and the antioxidant enzymes for successful germination using mutant seeds. In the *Arabidopsis cat2-1* mutant, intracellular H_2O_2 and redox perturbation were increased. In case of the *vte1-1* mutant lacking a gene that encodes the tocopherol cyclase, an increase in the redox active biosynthetic intermediate was observed [29]. Tocopherol is generally involved in the protection of lipids from oxidation. Tocopherols, also called vitamin E, functions as terminators of a PUFA recyclable chain reaction. The tocopheroxyl radical can be recycled back to the tocopherol by the reaction of ascorbate or other antioxidants [43]. The lack of the *vte1-1* function in the seeds releases lipid peroxy radicals. Although the *cat2-1* and *vte1-1* affect the seed germination to a certain level, plants lacking the *rbohD* gene can successfully complete germination (with time delay). The *rbohD* gene is involved in the conversion of the O_2 and NADPH to form superoxide and plays a vital role in the cell-to-cell propagation of the ROS and generation of the •OH. The •OH is essential for the cell wall loosening of endosperm [29].

3.6. Nonenzymatic antioxidants

Nonenzymatic antioxidants that actively participate in the ROS equilibrium are ascorbate, glutathione, and preoxiredoxins [44–47]. Low moisture content decreases the molecular mobility and the accessibility of substrates for the catalysis of antioxidant enzymes [44]. Ascorbate plays a major role in the progression of cell cycle, cell growth, hormonal signaling pathways, and embryogenesis. Ascorbate content of the seed decreased the H_2O_2 by increasing the peroxiredoxins [45]. Nonenzymatic antioxidants also determine the protection of cells against the ROS, particularly at the desiccation stage. Tocopherol is involved in the prevention of membrane damages by the LPO during a prolonged seed storage [4]. Peroxiredoxins protect the nuclear integrity and prevent against the oxidative damages of DNA under high levels of •OH radicals [46]. Involvement of the ascorbate-glutathione cycle alone in the seeds could be another vast area, which needs to be discussed separately.

3.7. Interplay between ROS, GA, and ABA

It has been proven that an inhibitor of the ROS, sodium benzoate, decreases the germination rate of the seed [47]. Diphenylene iodonium (DPI), an inhibitor of NADPH oxidase, also affects the germination rate [29]. On the other hand, methylviologen, involved in the release of superoxide from the mitochondrial respiratory chain breaks the seed dormancy [10]. Capacity of seeds to germinate or remain dormant is determined by the two important phytohormones such as GA (dormancy release) and/or ABA (dormancy induction). Bailly et al. reported that GA and ABA are interlinked with the ROS and the scavenging capacity of antioxidant enzymes [19]. Generally, GA is mainly used for the dormancy release, while ABA induces the dormancy. The GA is involved in the stimulation of •OH production, especially in the radicle, and it also downregulates the enzymes involved in the ROS detoxification. Contrastingly, ABA inhibits the Fenton reaction, where the iron (II) is oxidized by the H_2O_2 to form the iron (III) and the release of •OH [38]. The processes of seed germination and dormancy are linked with ROS accumulation [48]. The productions of H_2O_2 in the sunflower are higher in the germinating seeds than the dormant seeds [5]. Similar results have been observed by comparing the dormant and nondormant seeds of many plants, such as *Arabidopsis* [29], *Triticum aestivum* [49] and *Pisum sativum* L. [50]. Moreover, the H_2O_2 stimulates the signaling cascade which induces the expression of specific genes [24]. In addition to H_2O_2, accumulation of other ROS species such as O_2^- and •OH has also been observed in various plant species [50–53]. Bazin et al. mentioned that the germination of sunflower seeds was associated with the mRNA oxidation. The oxidation level of mRNA was higher in dormant seeds as compared with the nondormant seeds [32]. Genes such as *GA3ox1* and *GA3ox2* are involved in the synthesis of active GA [54]. Ishibashi et al. observed the induction of H_2O_2 in the aleurone layer by the GA in *Hordeum vulgare*, whereas ABA suppresses the production of H_2O_2. Furthermore, exogenous addition of H_2O_2 degrades Slender1 (SLN1), a well-known repressor of GA. Due to the induction of α-amylase (α-amy) and GAMyb, we can consider that H_2O_2 acts as an antagonist to ABA [55]. Cross-talk of the ROS with the phytohormones was reported to be mediated by the influx and efflux of Ca^{2+} ions [56, 57]. Bethke and Jones stated that H_2O_2 was involved in the programmed cell death [58]. Contrastingly, ABA increased the tolerance against the PCD [59]. Recent report suggested that the NO is also involved as a signaling messenger during the seed germination and dormancy process [60].

3.8. Protection against pathogens

The release of the ROS in the seeds during the development period protects the seeds against pathogens. It also induces the systemic-acquired resistance (SAR) and PCD. Especially when the ROS is mobile toward the seed coats, aleurone layers, and embryonic axis, the attack of microorganism is prevented by the induction of SAR and PCD [19, 61]. Briefly, the plasma membrane NADPH oxidase produces O_2^-, which is converted into H_2O_2 by SOD during the infection. Subsequently, H_2O_2 induces a hypersensitive reaction which leads to PCD of the infected cell. Eventually, H_2O_2 can also directly affect the pathogens [62, 63]. The main categories of genes involved in the H_2O_2 induction are related to defense, transcription, signaling (e.g., phosphatases, kinases), and importantly ROS synthesis and degradation. Perturbation of endosperm for the radicle emergence leads to the induction of defense-related genes. It helps to protect the newly germinating seeds from the pathogens [64]. During the seed germination process, lower concentrations of the ROS are involved in the cell signaling process, whereas higher concentrations trigger the PCD to facilitate the radical protrusion [65].

3.9. Endosperm weakening

Proteolytic cleavage of cell wall polymers is induced by the hydrolases such as mannose, glucanase, and cellulose. The scission of polysaccharides is a vital step to determine the radicle emergence. According to Muller et al., •OH accumulation is the main factor, which influences endosperm weakening by the breakdown of H-bonds in the cell wall-polysaccharide required for the radicle protrusion. Generally, •OH is extremely reactive and is considered as the most aggressive form of oxygenated derivatives [26]. Uncontrolled ROS accumulation affects the

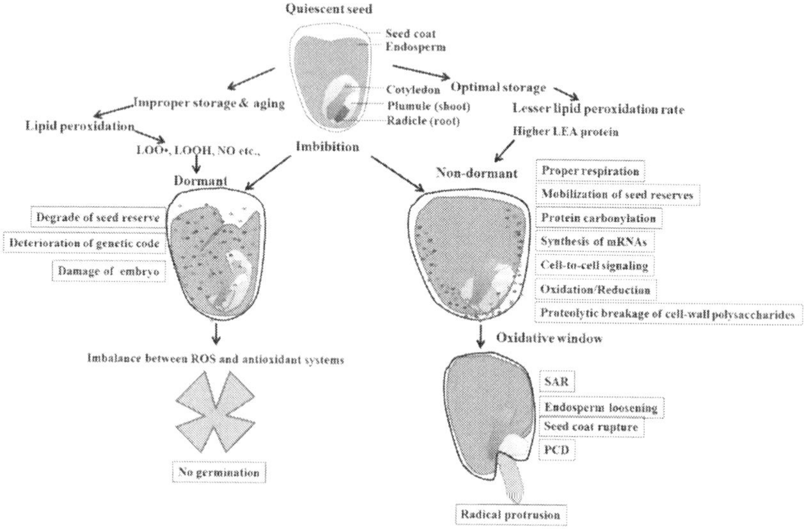

Figure 1. Schematic representation of the involvement of the reactive oxygen species (ROS) in the quiescent seed for the nondormancy and dormancy conditions. Dots represent the accumulation of ROS, blue for superoxide (O_2^-), brown for hydrogen peroxide (H_2O_2), and purple for hydroxyl radical (•OH). The LEA, late embryogenesis-abundant proteins; SAR, systemic acquired resistance; PCD, programmed cell death; LOO•, lipid peroxyl radical; LOOH, lipid hydroperoxide.

integrity of DNA, causing changes at the sequence level that impair proper seed germination and could be able to change the genetic code of the seeds. The O_2^- and H_2O_2 seem to be less reactive toward nucleic acid as compared to •OH [38]. However, cellular dysfunctions caused by ROS accumulation can be prevented by the antioxidants [40].

Oxidative damages caused by excess ROS are irreversible. The progressive conditional, temporal, and spatial distribution of the ROS tightly controlled by the antioxidant system leading to seed germination are also defined as 'oxidation window' (**Figure 1**).

4. Molecular Network of Gibberellins and Abscisic Acid in Germination/ Dormancy

As mentioned earlier, germination and dormancy release of seeds are governed by the GAs and ABA. According to previous reports, although there are many factors involved in seed germination, GA and ABA biosyntheses have been considered as an important internal factor for the release as well as the induction of dormancy [9]. Metabolisms of ABA and GA is always interrelated [2]. Seo et al. reported that *Arabidopsis aba2-2* (ABA inducer) mutant showed a higher level of GA biosynthesis as compared to the wild type. Contrastingly in the *gai* (GA insensitive) mutant, i.e., GA repressor, synthesis of ABA was higher, and also the degradation of ABA was lesser as than the wild type [11]. Antagonistic effect of GA and ABA is also cross-linked with other phytohormones such as ethylene, brassinosteriods, and auxins [9].

4.1. Dormancy breakage

Gibberellins are majorly responsible for the breakdown of dormancy [54]. However, in the later phase of embryogenesis, i.e., during the maturation drying, GA production must be reduced and ABA synthesis is upregulated for the proper maturation and to preserve seed vigor. During the imbibition stage, the endogenous GA_1 is released from the viable nondormant embryo to its endosperm. It increases the activities of several enzymes, such as amylase, proteases, ribonucleases, and β-glucanase, which induce the hydrolytic cleavage in the aleurone layer [66]. Moreover, hydrolytic enzymes are also involved in the transcription, transportation of metabolites, and PCD. Along with the GA biosynthesis, genes encoding for various functions are either overlapped or attributed toward the germination, and this process is controlled by the GA [66, 67]. Moreover, genes encoding gibberellin 3-oxidase (*GA3ox*) (GA biosynthesis) and the soluble GA receptor (*GA-insensitive dwarf, GID*) are vital for the induction of seed germination. During the germination, the *GA3ox2* is involved in the fast phase of GA synthesis by catalyzing the conversion of GA from an inactive to a bioactive form. In the GA synthesis, genes, such as *gibberellin 3-β-dioxygenase 1* (*GA3ox1*) and *gibberellin 20-β-dioxygenase 1* (*GA20ox1*), play as the positive regulators and *GA 2-oxidase*, especially *GA2ox1*, is involved in the negative regulation [67]. The *GA2ox1* gene is engaged with the catabolism of GA. The *Leafy cotyledon 1 (LEC1)* encodes for the katanin p60 subunit, which promotes the cell elongation in the microfibrill. Kroj et al. reported that the *lec2* and *fusca3* (*fus3*) directly influence the expression of the *GA3ox2* by binding on its *2S3* promoter region, particularly on the RY *cis-*

element motif. The RY motif also regulates the expression of genes encoding seed storage proteins (*SSP*) such as 2S albumins and 12S globulins. Four genes, such as *abi3* (ABA insensitive), *lec1*, *lec2* and *fus3*, are responsible for the conditional dormancy of embryos and are regulated by the ABA for proper maturation [68].

Higher expressions of cell wall-remodeling enzymes (CWRE) is associated with the radicle protrusion. Endo-β-mannanase, β-1,3-glucanase, expansin, xyloglucan endo-transglycosylase, pectin methylesterase, polygalacturonase, and galactanase are the notable major enzymes involved in the cell-wall modification [12]. In *Arabidopsis*, the application of GA induced the expression of the *extension-like gene 1* (*epr1*). The *epr1* is involved in the strengthening of micropylar endosperm cell wall to elongate the radicle of seeds [69]. Cell-wall-associated gene expression under the imbibition is preferentially linked with ABA and GAs in the endosperm weakening. During the rehydration, increased oil bodies and protein storage vacuole packed in cells start to mobilize from the endosperm cells to radicle tip. Penfeild et al. reported that lipids and proteins produced in the endosperm following the imbibition are higher than those produced in the embryo [70]. Consequently, the proteolytic activities are also at higher rates. The GA suppresses the activation of DELLA proteins to enhance the process of seed germination [66, 67, 71]. The E3 ubiquitin ligase SCF (skp1-cullin-F-box-Rbx) complex was reported to be involved in the degradation of DELLA proteins via the ubiquitin-26S proteasome pathway [72]. Perception of GA and its signal transduction determines various other mechanisms in the plant along with the seed germination (**Figure 2**).

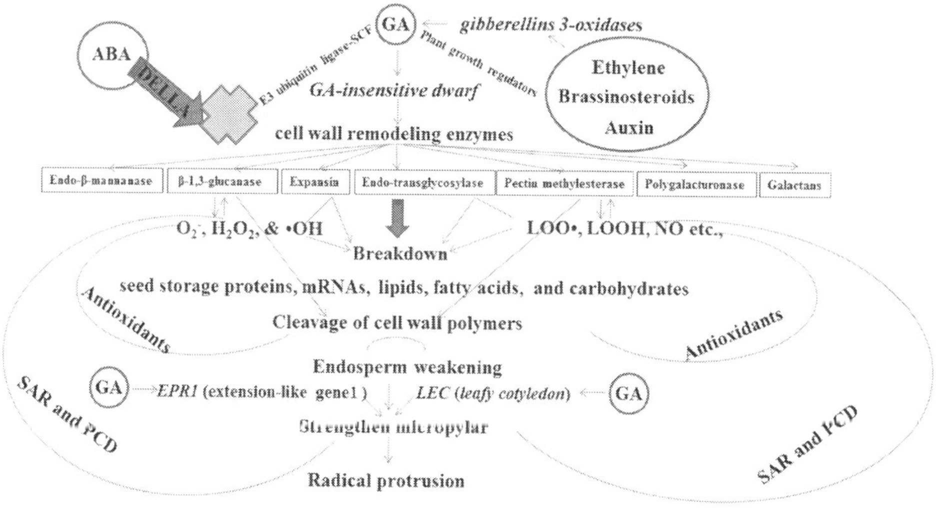

Figure 2. Simplified framework of the dormancy release mediated by GA in concordant with the ROS. Interactions between the genes are described in the text.

4.2. Dormancy induction

Enhanced dormancy occurred when the ABA content was increased in the *Arabidopsis* seeds due to the overexpression of ABA biosynthesis genes [73]. *Abscisic acid biosynthesis gene* (*aba2*), *9-cis-epoxycarotenoid dioxygenase* (*NCEDs*), *Glc-insensitive1* (*GIN1*), and *short-chain dehydrogenase/reductase1* (*SDR1*) are the major genes regulating the synthesis of ABA [74]. Increase in the seed dormancy is also associated with the expression levels of the gene *delay of germination1* (*DOG1*). Among the four regulatory genes, *abi3* and *fus3* are able to form the feedback loops. The *Lec1* and *lec2* mutants failed to express the *abi3* and *fus3*, suggesting the strong underlying molecular network between the regulators [75].

The ABA was involved in the vacuolation process to inhibit the endosperm loosening rather than the lipid breakdown [70]. Recently, it was found that the loss of function of the gene coding for E3 ligase ABI3-interacting protein 2 (AIP2) and ABI3-binding factor protein (AFP) leads to the ABA insensitivity and the nondormant phenotype even in the presence of ABA. Major receptors of the ABA are pyrabactin resistance1 (PYR1), PYR1-like protein (PYL), and regulatory components of ABA receptors (RCAR). Those loci encoding the main players in the ABA metabolism are also associated with the RNA translation and metabolism, protein-degradation pathways, and phosphatase components of the signaling pathways [69–75].

4.3. DELLA proteins

The DELLA proteins [GA insensitive (GAI), repressor of ga1-3 (RGA), and RGA-like proteins (RGL 1-3)] belong to the subfamily of plant-specific Glycyl-TRNA synthetase (GRAS) proteins. The name of GRAS proteins was derived from the initially identified members such as *GAI*, *RGA*, and SCARECROW (*SCR*) [76]. Potential of GA was repressed by the DELLA-domain containing members such as GAI, RGA, and RGL 1-3. Among them RGL2 has the major influence on seed dormancy. Most of DELLA proteins act as the repressors of the GA biosynthesis [55]. Seeds of DELLA mutants show the symptoms of irregular shape and proportion, especially in the protruded radicle [77].

Recent research found that the GA signal inactivated the functional domain of DELLA protein. The GA induced repression of the RGA through protein degradation rather than the blocking of translational process [71]. The deletion of the conserved motif VHYNP present in the DELLA region or the region between the VHYNP-DELLA, releases the dormancy by enhancing GA metabolism. Additionally, GA-dependent degradation of proteins is also associated with *SLENDER RICE 1* (*SLR1*), especially S/T/V, a regulatory region. However, in the wild type the function of the *SLR1* was not clearly distinct from the mutant. On the other hand, phosphorylation of SCF-E3 ligase and the ubiquitin-26S leads to the proteasome pathway of *SLEEPY1* (*SLY1*) and *GA-INSENSITIVE DWARF2* (*GID2*). Bioactive GA mainly focuses on the inactivation of RGA, GAI, and RGL 1-3 during the seed germination process [78, 79].

4.4. *DWARF, SLEEPY, PICKLE, SPY,* and *SECRET AGENT*

The presence of GA at higher levels makes the plant thin and watery. Contrastingly, lack of GA-biosynthetic genes or lesser amount of endogenous GA produces the thicker leaves with dwarf shoots. It was previously reported that GA-unresponsive dwarf phenotypes were observed in mutants such as *drwaf1* (*d1*) and *gid2* in rice, and *sly1* in *Arabidopsis* [80, 81]. The *GID1* and *SLY1* encoding for the F-box proteins are the subunit of the SCF complex belonging to the E3 ligase [78]. Genes (*SLR/RGA*) involved in the repression of GA were controlled by *GID2* and *SLY1* [82].

The *PICKLE* (*PKL*), an important positive regulator involved in the control of radicle protrusion, encodes a CHD3 chromatin remodeling factor. The *PKL* was reported to be actively involved in the later stage of seed germination, particularly on the root differentiation. Additionally, expression of the *PKL* was reported to be higher under an abiotic stress condition [83]. The gene *SPINDLY* (*SPY*) encodes the *O*-linked-*N*-acetyl glucosamine transferees (OGT), which negatively regulates the GA signaling (*GA1-3*) pathway [84]. The *SPY* decreases the GA effect by the suppression of the *SLR1* [85]. It is worthy mentioning that *GA1-3*, a precursor, is involved in the first step of the GA synthesis. Another gene, *SECRET AGENT* (*SEC*) which also encodes the *OGT* gene, found in *Arabidopsis*, did not show obvious phenotype alteration in the *sec* mutant (single mutant), whereas the mutant with both *spy* and *sec* (double mutant) showed the lethal effect in the gamete and also deeply affected the seed development. This double mutant could result from the alteration of not only the GA, but also the cytokine pathway [86]. The increased expression of tetratrico peptide repeats (TPRs) in *Arabidopsis* and *Petunia*, resulted with the repression of the *SPY*. The TRPs could either block directly by forming the inactive heterodimers or indirectly via proteins interacting with the *SPY* [87].

4.5. *AMYLASE*, *GAMYB*, and *LEAFY*

The enzyme amylase plays an important role in the hydrolysis of endosperm starch into usable sugars. This provides the necessary energy for the emergence of radicle. Plants possess both alpha (α)- and beta (β)-amylases. The expression of α-amylase in the aleurone layer is induced by GA. Activation of the *α-amy1* gene is mediated by GA-responsive elements (GARE) along with the C/TCTTTT and TATCCAT [66, 78]. For the *α-amy2* gene, along with the factors required for *α-amy1*, the BOX1/O2S-like elements are required [88]. The KGM, a Ser-Thr kinase, could repress the *α-amy1* by blocking the expression of the *HyGAMYB* [89]. Translation and stability of the *GAMYB* plays a major role for GA signaling. Meanwhile, interaction of novel zinc finger protein HRT (***H** ordeum ordeum* **r**epressor of **t**ranscription) with the *GARE* is able to repress the *α-amy2* gene expression [90]. After the inhibition for 12 h with GA_4, the *Arabidopsis ga1-3* mutant showed that 138 genes were upregulated and 120 genes were downregulated. The 20% of the upregulated genes possessed the TAACAAA-like sequences, indicating the importance of GARE in the cleavage of endosperm [91]. The *LEAFY* genes in the shoot apex are linked with the *GAMYB*-like genes. The *GAMYB* gene is also present in the anthers and expressed on the epidermis, endothecium, middle layer, and tapetum in the initial stages of development [92]. The GA activates the Ca^{2+} signaling for the synthesis of hydrolases. Decrease in the suppression of the *SLENDER1* (*SLN1*) increased the cytosolic Ca^{2+} level. The ABA inhibits the hydrolase by blocking the *sln1*, which directly affects the α-amy. By increasing the Ca^{2+}, GA activates the hydrolases via calmodulin signaling for successful emergence of the radicle [57–59, 82] (**Figure 3**).

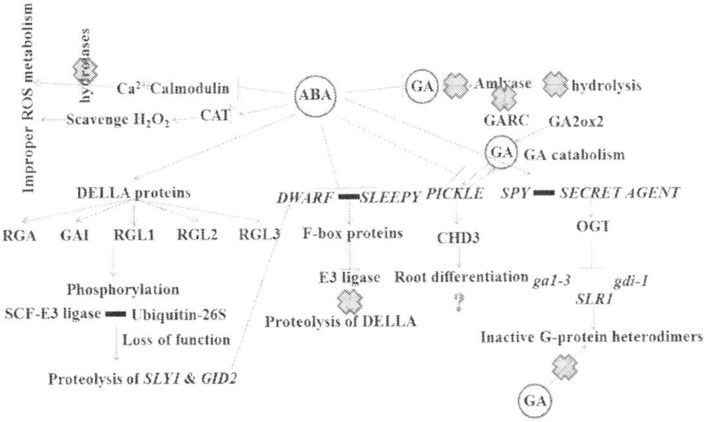

Figure 3. Heuristic networks on the dormancy induction by ABA. Interactions between the genes are described in the text. Arrows represent induction, barred lines represent inhibition, boxes represent sharing similar functions, question marks represent not clear, and crosses represent blocked or reduced synthesis. ABA, abscisic acid; Ca^{2+}, calcium; CAT, catalase; *GAI, GA insensitive; RGA, repressor of ga1-3; RGL, RGA-like protein; GID, gibberellin-insensitive dwarf; SPY, spindly; OGT, O-linked N-acetylglucosamine trasnferase* and CHD3 functions as chromatin remodeling factor.

5. Role of Other Phytohormones in Seed Germination/Dormancy

5.1. Ethylene

Ethylene was reported to be involved in various metabolisms such as dormancy breakage, root induction, defense against pathogens, and signaling [93]. Precisely, 1-aminocyclopropane-1-carboxylate oxidase (ACC oxidase), a precursor of ethylene synthesis, is required for higher synthesis of H_2O_2. Mutant lacking the GA synthesis gene (*gal-1*) possessed lower ethylene levels. Meanwhile, the exogenous supply of ethylene induced seed germination. The *ETR1* (ethylene receptor) was reported to be involved in the activation of serine/threonine kinase to overcome the dormancy in the *abi1-1* mutant [94]. In the straightway, mutants of *ein2* (*ethylene insensitive2*) and *etr1* (*ethylene receptor1*) showed higher degree of dormancy than the wild type. Especially, increase of dormancy ratio in the *ein2* mutant is directly proportional to the increased ABA content in the seeds. Meanwhile, decrease in the root ethylene content also decreased the ABA level [95].

5.2. Brassinosteroids

Brassinosteroids are well known for their functions in the cell elongation, cell cycle, and various other metabolisms. They are involved in the enhanced expression of *GA5*, a GA biosynthesis gene [96]. In the mutant of GA biosynthetic gene, *ga1*, and GA-insensitive mutant, *sly1*, application of brassinosteroids partially improved the seed germination under a lightdeprived condition [97]. Meanwhile, brassinosteroid receptor mutants such as *det2* (*deettiolated 2*) and *bri1* (*brassinosteroid insensitive*) were more sensitive to the ABA. Consequently, synthesis of GAs was also deeply affected [98]. This result signifies the importance of brassinosteriods in the GA synthesis for successful seed germination.

5.3. Auxin

Auxins are generally known for their roles in the root induction. Ogawa et al. reported upregulation of a number of auxin biosynthetic genes and genes encoding for auxin-carrying proteins in response to exogenous GA_4 application [91]. The GA was well known to promote the auxin synthesis and the transportation of ethylene. Chiwocha et al. (2005) evidenced that the interaction of ethylene biosynthetic genes with the auxin signaling genes such as *axr1* and *axr2* was mediated by GA [99]. The BIG-gene, named due to its large size, encodes the calossin/pushover protein involved in the efflux transportation of auxin [100]. Repressor of RGA proteins by the GA can be delayed by the attenuating auxin transportation or signaling [101]. Contrastingly, recent study in *Arabidopsis* by Lui et al. observed that the mutants of auxin receptors or biosynthesis genes showed the dramatic release of seed dormancy. This auxin-mediated seed dormancy was coordinated with ABA signaling [102]. Both GA and ABA have strong influence on auxins during germination and dormancy, respectively. This kind of cross-talk between the hormones helps in the flexibility of the embryo/seeds in response to the environmental stimuli.

6. Conclusions

During the developmental stage of embryos into the vigorous photoautrotropic organisms, numerous metabolic processes are activated and they include oxidation of proteins, cellular structural changes, and synthesis of macromolecules. The cascade of metabolic process ceases with the development of the radicle governed by the well-directed ROS accumulation. Interlinked relation between the GA and ABA aids in the proper development of the embryo, seed filling, desiccation tolerance, imbibition, hydrolysis, temporal and spatial distribution of ROS, proteolysis, and radicle protrusion. The recent evidences suggest that ABA-GA cross-talk with other phytohormones, such as ethylene, brassinosteroids and auxin, could play a vital role in the development of the seed. The important components other than the free radicals such as O_2^-, H_2O_2, and •OH pertaining to the seed potential is the NO. Tapping of the NO linked with the GA-ABA and their responses to the light and temperature could be one of the interesting areas getting more attention on the seed research.

Acknowledgements

Prabhakaran Soundararajan and Abinaya Manivannan were supported by a scholarship from the BK21 Plus Program, the Ministry of Education, Republic of Korea.

Author details

Prabhakaran Soundararajan[1], Abinaya Manivannan[1] and Byoung Ryong Jeong[1,2,3*]

*Address all correspondence to: brjeong@gmail.com

1 Division of Applied Life Science (BK21 Plus), Graduate School, Gyeongsang National University, Jinju, Republic of Korea

2 Institute of Agriculture and Life Science, Gyeongsang National University, Jinju, Republic of Korea

3 Research Institute of Life Science, Gyeongsang National University, Jinju, Republic of Korea

References

[1] Bewley JD, Black M. Seeds: Physiology of Development and Germination. 3rd ed. New York: Plenum Press; 1994.

[2] Bewley JD. Seed germination and dormancy. Plant Cell. 1997; 9:1055–1066.

[3] Lang GA. Plant Dormancy. Oxford, UK: CAB International; 1996.

[4] Bailly C. Active oxygen species and antioxidants in seed biology. Seed Science Research. 2004; 14:93–107.

[5] El-Maarouf-Bouteau H, Bailly C. Oxidative signaling seed germination and dormancy. Plant Signaling and Behavior. 2008; 3:175–182.

[6] Considine MJ, Foyer CH. Redox regulation of plant development. Antioxidants and Redox Signaling. 2014; 21:1305–1326.

[7] Koornneef M, Karssen CM. Seed dormancy and germination. In: Meyerowitz EM, Somerville CR, editors. Arabidopsis. Cold Spring Harbor, NY: Cold Spring Harbor Laboratory Press; 1994. pp. 313–334.

[8] Karssen CM. Hormonal regulation of seed development, dormancy, and germination studied by genetic control. In: Kigel J, Galili J, editors. Seed Development and Germination. New York: Marcel Dekker; 1995. pp. 333–350.

[9] Brady SM, McCourt P. Hormone cross-talk in seed dormancy. Journal of Plant Growth Regulation. 2003; 22:25–31.

[10] Bahin E, Bailly C, Sotta B, Kranner I, Corbineau F, Leymarie J. Crosstalk between reactive oxygen species and hormonal signalling pathways regulates grain dormancy in barley. Plant Cell Environment. 2011; 34:980–993.

[11] Seo M, Hanada A, Kuwahara A, Endo A, Okamoto M, Yamauchi Y, North H, Marion-Poll A, Sun TP, Koshiba T, et al. Regulation of hormone metabolism in Arabidopsis seeds: phytochrome regulation of abscisic acid metabolism and abscisic acid regulation of gibberellins metabolism. Plant Journal. 2006; 48:354–366.

[12] Kucera B, Cohn MA, Leubner-Metzger G. Plant hormone interactions during seed dormancy release and germination. Seed Science Research. 2005; 15:281–307.

[13] Law SR, Narsai R, Taylor NL, Delannoy E, Carrie C, Giraud E, Millar AH, Small I, Whelan J. Nucleotide and RNA metabolism prime translational initiation in the earliest events of mitochondrial biogenesis during *Arabidopsis* germination. Plant Physiology. 2012; 158:1610–1627.

[14] Law SR, Narsai R, Whelan J. Mitochondrial biogenesis in plants during seed germination. Mitochondrion. 2014; 19:214–221.

[15] Meitha K, Konnerup D, Colmer TD, Considine JA, Foyer CH, Considine MJ. Spatiotemporal relief from hypoxia and production of reactive oxygen species during bud burst in grapevine (*Vitis vinifera*). Annals of Botany. 2015.

[16] Halliwell B, Gutteridge JMC. Free Radicals in Biology and Medicine. 3rd ed. New York: Oxford University Press; 1999.

[17] Cui K, Xing G, Liu X, Xing G, Wang Y. Effect of hydrogen peroxide on somatic embryogenesis of *Lycium barbarum* L. Plant Science. 1999; 146:9–16.

[18] Kärkönen A, Kuchitsu K. Reactive oxygen species in cell wall metabolism and development in plants. Phytochemistry. 2015; 112:22–32.

[19] Bailly C, El-Maarouf-Bouteau H, Corbineau F. From intracellular signaling networks to cell death: the dual role of reactive oxygen species in seed physiology. Comptes Rendus Biologies. 2008; 331:806–814.

[20] Beckman KB, Ames BN. Oxidants, antioxidants, and aging. In: Scandalios JG, editor. Oxidative Stress and the Molecular Biology of Antioxidant Defenses. New York: Cold Spring Harbor Laboratory Press; 1997. pp. 201–246.

[21] Nakabayashi K, Okamoto M, Koshiba T, Kamiya Y, Nambara E. Genome-wide profiling of stored mRNA in Arabidopsis thaliana seed germination: epigenetic and genetic regulation of transcription in seed. The Plant Journal. 2005; 41:697–709.

[22] Han B, Hughes DW, Galau GA, Bewley JD, Kermode AR. Changes in late embryogenesis abundant (LEA) messenger RNAs and dehydrins during maturation and premature drying of *Ricinus communis* L. seeds. Planta. 1996; 201:27–35.

[23] Jiang L, Kermode AR. Role of desiccation in the termination of expression of genes for storage proteins. Seed Science Research. 1994; 4:149–173.

[24] Foyer CH, Noctor G. Redox sensing and signalling associated with reactive oxygen in chloroplasts, peroxisomes and mitochondria. Physiologia Plantarum. 2003; 119:355–364.

[25] Corpas FJ, Barroso JB, del Rio LA. Peroxisomes as a source of reactive oxygen species and nitric oxide signal molecules in plant cells. Trends in Plant Science. 2001; 6:145–150.

[26] Müller K, Heß B, Leubner-Metzger G. A role for reactive oxygen species in endosperm weakening. Seeds: Biology, Development and Ecology. 2007; 2:87–95.

[27] Desikan R, Mackerness SAH, Hancock JT, Neill SJ. Regulation of the *Arabidopsis* transcriptome by oxidative stress. Plant Physiology. 2001; 127:159–172.

[28] Walters C. Understanding the mechanisms and kinetics of seed aging. Seed Science Research. 1998; 8:223–244.

[29] Leymarie J, Vitkauskaité G, Hoang HH, Gendreau E, Chazoule V, Meimoun P, Corbineau F, El-Maarouf-Bouteau H, Bailly C. Role of reactive oxygen species in the regulation of Arabidopsis seed dormancy. Plant and Cell Physiology 2012; 1:96–106.

[30] Dietz K, Jacquot J, Harris G. Hubs and bottlenecks in plant molecular signalling networks. New Phytology. 2010; 188:919–938.

[31] Oracz K, El-Maarouf-Bouteau H, Kranner I, Bogatek R, Corbineau F, Bailly C. The mechanisms involved in seed dormancy alleviation by hydrogen cyanide unravel the role of reactive oxygen species as key factors of cellular signaling during germination. Plant Physiology. 2009; 150:494–505.

[32] Bazin J, Langlade N, Vincourt P, Arribat S, Balzergue S, El-Maarouf-Bouteau H, et al. Targeted mRNA oxidation regulates sunflower seed dormancy alleviation during dry after-ripening. The Plant Cell. 2011; 23:2196–2208.

[33] Chandrakuntal K, Shah AK, Thomas NM, Karthika V, Laloraya M, Kumar PG, et al. Blue light exposure targets NADPH oxidase to plasma membrane and nucleus in wheat coleoptiles. Journal of Plant Physiology. 2010; 29:232–241.

[34] Nyström, T. Role of oxidative carbonylation in protein quality control and senescence. The EMBO Journal. 2005; 24:1311–1317.

[35] Wong CM, Marcocci L, Liu LL, Suzuk YJ. Cell signaling by protein carbonylation and decarbonylation. Antioxidants and Redox Signaling. 2010; 12:393–404.

[36] Barba-Espín G, Diaz-Vivancos P, Job D, Belghazi M, Job C, Hernandes JA. Understanding the role of H_2O_2 during pea seed germination: a combined proteomic and hormone profiling approach. Plant Cell & Environment. 2011; 34:1907–1919.

[37] Job C, Rajjou L, Lovigny Y, Belghazi M, Job D. Patterns of protein oxidation in Arabidopsis seeds and during germination. 2005; 138:790–802.

[38] Muller K, Carstens AC, Linkies A, Torres MA, Leubner-Metzger G. The NADPH-oxidase *AtrbohB* plays a role in Arabidopsis seed after-ripening. New Phytology. 2009; 184:885–897.

[39] Prasad TK, Anderson, MD, Martin, BA, Stewart, CR. Evidence for chilling-induced oxidative stress in maize seedlings and a regulatory role for hydrogen peroxide. The Plant Cell. 1994; 6:65–74.

[40] Pulido P, Cazalis R, Cejudo FJ. An antioxidant redox system in the nucleus of wheat seed cells suffering oxidative stress. The Plant Journal. 2009; 57:132–145.

[41] Queval G, Issakidis-Bourguet E, Hoeberichts FA, Vandorpe M, Gakie`re B, Vanacker H, et al. Conditional oxidative stress responses in the Arabidopsis photorespiratory mutant cat2 demonstrate that redox state is a key modulator of daylengthdependent gene expression, and define photoperiod as a crucial factor in the regulation of H_2O_2-induced cell death. The Plant Journal. 2007; 52:640–657.

[42] Guan LQM, Scandalios JG. Catalase gene expression in response to auxin-mediated developmental signals. Physiologia Plantarum. 2002; 114:288–295.

[43] Mene-Saffrane L, Jones AD, DellaPenna D. Plastochromanol-8 and tocopherols are essential lipid-soluble antioxidants during seed desiccation and quiescence in *Arabidopsis*. Proceedings of the National Academy of Sciences USA. 2010; 107:17815–17820.

[44] Bailly C, Benamar A, Corbineau F, Côme D. Changes in malondialdehyde content and in superoxide dismutase, catalase and glutathione reductase activities in sunflower seeds as related to deterioration during accelerated aging. Physiologia Plantarum. 1996; 97:104–110.

[45] De Tullio MC, Arrigoni O. The ascorbic acid system in seeds: to protect and to serve. Seed Science Research. 2003; 13:249–260.

[46] Aalen RB. Peroxiredoxin antioxidants in seed physiology. Seed Science Research. 1999; 9:285–295.

[47] Hung KT, Kao CH. Nitric oxide acts as an antioxidant and delays methyl jasmonate-induced senescence of rice leaves. Journal of Plant Physiology. 2004; 161:43–52.

[48] Gomes MP, Garcia QS. Reactive oxygen species and seed germination. Biologia. 2013; 68:351–357.

[49] Caliskan M, Cuming AC. Spatial specificity of H_2O_2 generating oxalate oxidase gene expression during wheat embryo germination. The Plant Journal. 1998; 15:165–171.

[50] Wojtyla Ł, Garnczarska M, Zalewski T, Bednarski W, Ratajczak L, Jurga S. A comparative study of water distribution, free radical production and activation of antioxidative metabolism in germinating pea seeds. Journal of Plant Physiology. 2006; 163:1207–1220.

[51] Gidrol X, Lin WS, Degousee N, Yip SF, Kush A. Accumulation of reactive oxygen species and oxidation of cytokinin in germinating soybean seeds. European Journal of Biochemistry. 1994; 224:21–28.

[52] Liu X, Xing D, Li L, Zhang L. Rapid determination of seed vigor based on the level of superoxide generation during early imbibition. Photochemical Photobiological Science. 2007; 6:767–774.

[53] Schopfer P, Plachy C, Frahry G. Release of reactive oxygen intermediates (superoxide radicals, hydrogen peroxide, and hydroxyl radicals) and peroxidase in germinating radish seeds controlled by light, gibberellin, and abscisic acid. Plant Physiology. 2001; 125:1591–1602.53.

[54] Yamaguchi S. Gibberellin metabolism and its regulation. Annual Review of Plant Biology. 2008; 59:225–251.

[55] Ishibashi Y, Tawaratsumida T, Kondo K, et al. Reactive oxygen species are involved in gibberellin/abscisic acid signaling in barley aleurone cells. Plant Physiology. 2012; 158:1705–1714.

[56] Bush DS, Biswas AK, Jones RL. Hormonal regulation of Ca^{2+} transport in the endomembrane system of the barley aleurone. Planta. 1993; 189:507–515.

[57] Bethke PC, Jones RL. Ca^{2+}-calmodulin modulates ion channel activity in storage protein vacuoles of barley aleurone cells. Plant Cell. 1994; 6:277–285.

[58] Bethke PC, Jones RL. Cell death of barley aleurone protoplasts is mediated by reactive oxygen species. Plant Journal. 2001; 25:19–29.

[59] Fath A, Bethke PC, Jones RL. Enzymes that scavenge reactive oxygen species are downregulated prior to gibberellic acid-induced programmed cell death in barley aleurone. Plant Physiology. 2001; 126:156–166.

[60] Sanz L, Albertos P, Mateos I, et al. Nitric oxide (NO) and phytohormones crosstalk during early plant development. Journal of Experimental Botany. 2015; 66:2857–2868.

[61] Stacy RAP, Murnthe E, Steinum T, Sharma B, Aalen RB. A peroxiredoxin antioxidant is encoded by a dormant- related gene, *Perl*, expressed during late development in the aleurone and embryo of barley grains. Plant Molecular Biology. 1996; 31:1205–1216.61.

[62] Kumar SJ, Prasad SR, Banerjee R, Thammineni C. Seed birth to death: dual functions of reactive oxygen species in seed physiology. Annals of Botany. 2015.

[63] Levine A, Tenkanen R, Dixon R, Lamb C. H_2O_2 from the oxidative burst orchestrates the plant hypersensitive disease resistance response. Cell. 1994; 79:583–593.

[64] Foyer CH, Noctor G. Redox homeostasis and antioxidant signalling: a metabolic interface between stress perception and physiological responses. Plant Cell. 2005; 17:1866–1875.

[65] Shaban M, Abasalt RA, Ghorban DM, Tahmine B. Review on dual role of reactive oxygen species in seed physiology and germination. International Journal of Agriculture and Crop Sciences. 2013; 5:2390–2393.

[66] Jacobsen JV, Gubler F, Chandler PM. Gibberellin action in germinating cereal grains. In: Davies PJ, editor. Plant Hormones: Physiology, Biochemistry, and Molecular Biology. Dordrecht: Kluwer Acad. 1995. pp. 164–193.

[67] Olszewski N, Sun TP, Gubler F. Gibberellin signaling: biosynthesis, catabolism, and response pathways. Plant Cell Supplement. 2002; S61–S80.

[68] Raz V, Bergervoet JHW, Koornneef M. Sequential steps for developmental arrest in Arabidopsis seeds. Development. 2001; 128:243–252.

[69] Dubreucq B, Berger N, Vincent E, Boisson M, Pelletier G, Caboche M, Lepiniec L. The *Arabidopsis AtEPR1* extensin-like gene is specifically expressed in endosperm during seed germination. The Plant Journal. 2000; 23:643–652.

[70] Penfield S, Rylott EL, Gilday AD, Graham S, Larson TR, Graham IA. Reserve mobilization in the *Arabidopsis* endosperm fuels hypocotyls elongation in the dark, is independent of abscisic acid, and requires phosphoenolpyruvate carboxykinase 1. Plant Cell. 2004; 16:2705–2718.

[71] Holdsworth MJ, Bentsink L, Soppe WJ. Molecular networks regulating Arabidopsis seed maturation, after-ripening, dormancy and germination. New Phytologist. 2008; 179(1):33–54.

[72] McGinnis KM, Thomas SG, Soule JD, Strader LC, Zale JM, Sun TP, Steber CM. The Arabidopsis SLEEPY1 gene encodes a putative F-box subunit of an SCF E3 ubiquitin ligase. Plant Cell. 2003; 15:1120–1130.

[73] Kushiro T, Okamoto M, Nakabayashi K, Yamagishi K, Kitamura S, Asami T, Hirai N, Koshiba T, Kamiya Y, Nambara E. The *Arabidopsis* cytochrome P450CYP707A encodes ABA 8′-hydroxylases: key enzymes in ABA catabolism. EMBO Journal. 2004; 23:1647–1656.

[74] Nambara E, Marion-Poll A. ABA action and interactions in seeds. Trends in Plant Science. 2003; 8:213–217.

[75] To A, Valon C, Savino G, Guilleminot J, Devic M, Giraudat J, Parcy F. A network of local and redundant gene regulation governs Arabidopsis seed maturation. Plant Cell. 2006; 18:1642–1651.

[76] Pysh LD, Wysocka Diller JW, Camilleri C, Bouchez D, Benfey PN. The GRAS gene family in Arabidopsis: sequence characterization and basic expression analysis of the SCARECROW-LIKE genes. The Plant Journal. 1999; 18:111–119.

[77] Cao DN, Hussain A, Cheng H, Peng JR. Loss of function of four DELLA genes leads to light- and gibberellin-independent seed germination in Arabidopsis. Planta. 2005; 223:105–113.

[78] Sun TP, Gubler F. Molecular mechanism of gibberellin signaling in plants. Annual Review of Plant Biology. 2004; 55:197–223.

[79] Penfield S, Gilday AD, Halliday KJ, Graham IA. DELLA-mediated cotyledon expansion breaks coat-imposed seed dormancy. Current Biology. 2006; 16:2366–2370.

[80] Mitsunaga S, Tashiro T, Yamaguchi J. Identification and characterization of gibberellin-insensitive mutants selected from among dwarf mutants of rice. Theoretical and Applied Genetics. 1994; 87:705–712.

[81] Steber CM, Cooney S, McCourt P. Isolation of the GA-response mutant sly1 as a suppressor of ABI1-1 in *Arabidopsis thaliana*. Genetics. 1998; 149:509–521.

[82] Vierstra RD. The ubiquitin/26S proteasome pathway, the complex last chapter in the life of many plant proteins. Trends in Plant Science. 2003; 8:135–142.

[83] Ogas J, Kaufmann S, Henderson J, Somerville C. PICKLE is a CHD3 chromatin-remodeling factor that regulates the transition from embryonic to vegetative development in Arabidopsis. Proceedings of the National Academy of Sciences USA. 1999; 96:13839–13844.

[84] Jacobsen SE, Binkowski KA, Olszewski NE. SPINDLY, a tetratricopeptide repeat protein involved in gibberellin signal transduction in Arabidopsis. Proceedings of the National Academy of Sciences USA. 1996; 93:92–96.

[85] Swain SM, Tseng T-S, Olszewski NE. Altered expression of SPINDLY affects gibberellin response and plant development. Plant Physiology. 2001; 126:1174–1185.

[86] Hartwek LM, Scott CL, Olszewski NE. Two O-Linked N-acetylglucosamine transferase genes of *Arabidopsis thaliana* L. Heynh. have overlapping functions necessary for gamete and seed development. Genetics. 2002; 161:1279–1291.

[87] Izhaki A, Swain SM, Tseng T-S, Borochov A, Olszewski NE, Weiss D. The role of SPY and its TPR domain in the regulation of gibberellin action throughout the life cycle of *Petunia hybrida* plants. The Plant Journal. 2001; 28:181–190.

[88] Gubler F, Kalla R, Roberts J, Jacobsen JV. Gibberellin-regulated expression of a myb gene in barley aleurone cells: evidence for Myb transactivation of ahigh-pI α-amylase gene promoter. Plant Cell. 1995; 7:1879–1891.

[89] Woodger FJ, Gubler F, Pogson BJ, Jacobsen JV. A Mak-like kinase is a repressor of GAMYB in barley aleurone. The Plant Journal. 2003; 33:707–717.

[90] Raventos D, Meier C, Mattsson O, Jensen AB, Mundy J. Fusion genetic analysis of gibberellin signaling mutants. The Plant Journal. 2000; 22:427–438.

[91] Ogawa M, Hanada A, Yamauchi Y, Kuwahara A, Kamiya Y, Yamaguchi S. Gibberellin biosynthesis and response during *Arabidopsis* seed germination. Plant Cell. 2003; 15:1591–1604.

[92] Gocal GFW, Sheldon CC, Gubler F, Moritz T, Bagnall DB, et al. GAMYB-like genes, flowering and gibberellins signaling in *Arabidopsis*. Plant Physiology. 2001; 127:1682–1693.

[93] Bleecker AB, Esch JJ, Hall AE, Rodriguez FI, Binder BM. The ethylene-receptor family from Arabidopsis: structure and function. Philosophical Transactions of the Royal Society of London B: Biological Science. 1998; 353:1405–1412.

[94] Beaudoin N, Serizet C, Gosti F, Giraudat J. Interactions between abscisic acid and ethylene signalling cascades. Plant Cell. 2000; 12:1103–1105.

[95] Ghassemian M, Nambara E, Cutler S, Kawaide H, Kamiya Y, McCourt P. Regulation of abscisic acid signalling by the ethylene response pathway in Arabidopsis. Plant Cell. 2000; 12:1117–1126.

[96] Bouquin T, Meier C, Foster R, Nielse ME, Mundy J. Control of specific gene expression by gibberellin and brassinosteroid. Plant Physiology. 2001; 127:450–458.

[97] Clouse SD, Sasse JM. Brassinosteroids: essential regulators of plant growth and development. Annual Review of Plant Physiology and Plant Molecular Biology. 1998; 49:427–451.

[98] Steber CM, McCourt P. A role for brassinosteroids in germination in Arabidopsis. Plant Physiology. 2001; 125:763–769.

[99] Chiwocha SD, Cutler AJ, Abrams SR, Ambrose SJ, Yang J, Ross AR, Kermode AR. The *etr1-2* mutation in Arabidopsis thaliana affects the abscisic acid, auxin, cytokinin and gibberellin metabolic pathways during maintenance of seed dormancy, moist-chilling and germination. The Plant Journal. 2005; 42:35–48.

[100] Kanyuka K, Praekelt U, Franklin KA, Billingham OE, Hooley R, et al. Mutations in the huge Arabidopsis gene BIG affect a range of hormone and light responses. The Plant Journal. 2003; 35:57–70.

[101] Fu X, Harberd NP. Auxin promotes Arabidopsis root growth by modulating gibberellin response. Nature. 2003; 421:740–743.

[102] Liu X, Zhang H, Zhao Y, Feng Z, Li Q, Yang HQ, Luan S, Li JM, He ZH. Auxin controls seed dormancy through stimulation of abscisic acid signaling by inducing ARF-mediated ABI3 activation in Arabidopsis. Proceedings of the National Academy of Sciences. 2013; 110:15485–15490.

Chapter 4

Effects of Some Hormone Applications on Germination and Morphological Characters of Endangered Plant Species *Lilium artvinense* L. Seeds

Kerim Guney, Mehmet Cetin, Hakan Sevik and
Kudret Betül Güney

Additional information is available at the end of the chapter

http://dx.doi.org/10.5772/64466

Abstract

Lilies are economically important plants because of their large and attractive flowers. Thus, many wild species of lilies have been cultivated to produce Lilium bulbs or flowers. This work was conducted to analyse the effect of hormone applications on Lilium artvinense (Syn: Lilium ponticum K. Koch., Lilium ponticum var. artvinense (Miscz.) P. H. Davis and D. M. Hend., Lilium carniolicum var. artvinense (Miscz.) P. H. Davis and D. M. Hend and Lilium pyrenaicum var. artvinense (Miscz.) V.A. Matthews) seeds on germination percentage and seedlings morphological traits. In the research, 1000, 3000 and 5000 ppm doses of IAA, IBA, NAA and GA3 hormones were applied to L. artvinense seeds and approximately 180 days later, the number of roots, root length, offset stem height and diameter were assessed. As a result, while the control group except 5000 ppm NNA application achieved an increase in the percentage of germination (40%) of all the applications. Germination frequency up to 100% was obtained using 5000 ppm GA3. Effects of hormone applications on other key morphological characters (rooting percentage, root height, number of scions, scion height and width) are described in terms of growth rate between 1.27 and 2.44.

Keywords: *Lilium artvinense*, hormone, seed, germination, morphological characters, endangered species, root height, scion height, scion width

1. Introduction

In 2000, 47% of world population (2.9 billion people) was living in urban areas, while it is expected that 60–90% of the world's population will be living in urban areas by 2030 [1, 2]. In European countries, more than two-thirds of the whole population are living in urban areas [2, 3].

Rapid urbanization and industrialization caused human beings to distance themselves from nature more and more with each passing day and disrupted humanity's harmony with landscape. As a part of the natural world, humans have brought a piece of universe to every single place they have lived—be it a houseplant, a small garden or a carefully organized park [4]. In today's modern life, vegetation is recognized as an indicator of life quality and livability in cities [5].

In the place they grow, plants reduce air pollution [6–10], reduce noise [11], increase aesthetic value [12], have a positive psychological effect [13, 14], provide energy conservation [15], prevent erosion [16], reduce wind speed and hold the soil with their roots, thus preventing washing away of the soil with rainfalls and streams, and protect wildlife and hunting resources. Open-green areas with plantation are important activity areas for both adults and children [17, 18]. Besides, indoor plants increase the productivity of people working in the environment they grow in [19] as well as relieving physiological stress and reducing negative emotions [20–22].

Such functional use of plants led to the development of ornamental plants market, which finally has reached a point which is economically significant. In 145 countries around the world, the cultivation of ornamental plants is carried out on a total area of 220,000 ha, and the trade volume of ornamental plants is around $50 billion USD [23].

Such advancement of the ornamental plants market made the researchers to be interested in various issues such as defining the distribution areas [24, 25], protection [26–30], cultivation [31, 32], resistance to stress factors [33, 34], various applications [11, 35, 36], genetic variability [16, 37–40], their relationship with the environment [41–44], raising awareness about plants and legal dimension of the issue [45–47], thus resulting in various works on these issues. In addition to these, a large number of reports were conducted on especially the generative [48–55] and vegetative cultivation methods, and the studies are still going on [56]. However, it is of particular importance to define new species, variations, forms, cultivars or hybrids that can be launched especially to the ornamental plants market, as all plants are of high economic importance.

One of the species that can be dealt with in the ornamental plants market is *Lilium*. *Lilium* belongs to tribe *Lilieae* of *Liliaceae* and contains about 120 species [57]. Lily is one of the most important horticultural plants, and many wild species and cultivars have been cultivated for bulb and ornamental flower production [58]. Thus, many wild species of lilies have been cultivated to produce *Lilium* bulbs or flowers. Almost all cultivars of lilies are used in ornamental flower. Especially Asiatic and oriental hybrids receive a great deal of attention in the international market [59]. Due to the demand for *Lilium* in recent years, the production of *Lilium* has rapidly increased. For example, *Lilium* was being grown in the Netherlands in an area of 102 ha in 1960, which rose to 2412 ha in 1990 [60, 61].

However, picking especially rare and endemic species in some countries damages the populations of the species in the natural world. The easiest and the most effective way of picking the species from the natural environment is the identification of methods for producing the species in nursery conditions. If such methods are easy, cheap and practical, they can be put into practice more effectively. Seed-based production is of vital importance so as not to damage the natural populations of endangered species in particular.

'IUCN Red Data BOOK' and 'National List of European Threatened Plants List' contain three *Lilium* species from Europe. *Lilium artvinense* (Syn: *Lilium carniolicum* var. *artvinense, Lilium ponticum* K. Koch., *Lilium ponticum* var. *artvinense* (Miscz.) P. H. Davis and D. M. Hend., *Lilium carniolicum* var. *artvinense* (Miscz.) P. H. Davis and D. M. Hend and *Lilium pyrenaicum* var. *artvinense* (Miscz.) V. A. Matthews) [62] is one of them [63]. Exposed over a limited area, *L. artvinense* draws attention with its yellow flowers. Its natural population is severely damaged when its bulbs are picked from the natural world. To protect the natural population of the species, easy and cheap means of producing this species should be taught to locals, thereby preventing them from picking the species from the natural environment

Previous research has mostly attempted to determine the microculture techniques of producing lilies. However, these techniques are difficult, expensive and impracticable for locals. This report examines the production of *L. artvinense* by the use of seeds.

The present review aims to define the effect of some plant growth regulators (hormones) on germination and certain morphological traits of *L. artvinense* seeds. Plant growth-regulating substances are organic chemical compounds, produced naturally in plants or applied externally, and that, even in small amounts, can have positive and negative effects on growth, development and other physiological functions, alone or in combination with other growth regulators. They can be effective in the tissue where they are synthesized, or transported to other parts of the plant and sustain this effectiveness in other organs of the plant [64, 65].

Since 1930s, plant growth regulators have been investigated for their function in agricultural products [64]. Today, there are many works with the aim to define the effects of external application of hormones commonly used to affect the growth rate and development of a plant from the stage of germination to the harvest and post-harvest treatment, and to determine proper type, concentration and time for the application of hormones. The use of hormones is so common today that they are investigated and used in agriculture, forestry and production of various plant species from ornamental to medicinal and aromatic ones [66–68].

The most relevant role of hormones is related to the control and coordination of cell division, growth and differentiation [69]. Plant hormones including ethylene, gibberellins, auxin (indole-3-acetic acid (IAA)), abscisic acid (ABA), brassinosteroids and cytokinins are biochemical substances controlling many biochemical processes and physiological in the plant [69–71]. These key molecules are produced not only by plants but also by soil microbes.

The hormones can be said to be mainly used to ensure propagation by cutting, to increase germinating power of seed, to promote and inhibit blooming, to increase plant resistance against cold, to increase seed formation in fruits, to increase fruit size, to extend fruit storage time, to increase plant resistance against diseases and pests, to control weeds, to prevent lodging in cotton and cereals, to prevent pre-harvest fruit drop, to synchronize maturity of all plants, to hasten maturity, to remove dormancy and to initiate the growth of suckers and tubers especially during the tissue culture studies [64].

Among all hormones, auxins and gibberellins are the most widely used hormones with usage rates of 20 and 17%, respectively [64]. Auxins, which were firstly used in 1929 [72], mostly cause the expansion and growth of cells and initiate cell elongation, tissue growth and root formation. Auxins are synthesized in all higher plant species, and the most commonly found auxin is indole-3-acetic acid [73].

Other commonly found auxins other than IAA are indole butyric acid (IBA), naphthalene acetic acid (NAA), naphthoxyacetic acid (NOAA), phenoxyacetic acid (POAA), 2,4-D, phenylacetic acid (PAA), para-chlorophenoxyacetic acid (4-CPA) and 2,4,5-trichlorophenoxyacetic acid (2,4,5-T) [64].

The oldest and most common use of growth-regulating hormones is to stimulate rooting process. Among the hormones in the auxin class, IAA, IBA and NAA are used for stimulating the rooting of cuttings. However, the most commonly applied hormone in agricultural practices is IBA [74]. IBA is applied as a highly concentrated (1000–8000 ppm) and dilute (10–250 ppm) solution. In achieving successful rooting, temperature, light conditions and water supply are also effective parameters [75].

The findings of this study are in conformity with other reports in the literature. Guney et al. [61] reported that hormone treatment caused a significant increase in the rooting percentage of *Lilium martagon* seeds, and the rooting percentage of 28.40% in the control groups could be increased with IAA, IBA, NAA and GA3 treatments. They found the highest rooting percentage in the seeds treated with IAA at 5000 ppm (86.6%).

Turhan [65] reported that the effect of hormone treatment on the rooting percentage of *L. martagon* was statistically significant. He stated that the highest rooting percentage was 66.67%, which was obtained with 3000 ppm IBA + 1000 ppm IAA treatment. In his report, the rooting percentage was found to be 61.73% with 3000 ppm IBA + 3000 ppm NAA treatment and 59.26% with 1000 ppm IBA + 1000 ppm GA3 treatment, while it was 45.83% in the control group. Sevik and Cetin [76] also reported that 3000 ppm IAA treatment in *L. artvinense* seeds increased the rooting percentage from 28.57 to 80.22%.

Rooting percentage can be said to be one of the most important morphological traits that have been examined in the researches. The main focus of the works conducted so far has been the treatments that can increase rooting percentage. The effects of auxins on rooting and plant growth have been so far widely demonstrated [77]. These hormones have been proven to increase rooting percentage in the seeds of *Robinia pseudoacacia* [78], *Pseudotsuga menziesii* [79], *Oryza sativa* [80], *Pisum sativum* [81] and many other plant species.

IBA is one of the most commonly used and studied hormones. Its impact on rooting is continuous and extremely high [82]. There are many works showing that IBA treatments alone are highly effective in increasing rooting percentage. Polat et al. [83] reported that the rooting percentage of 5% in the control group of plum cuttings increased up to 60% with 500 ppm IBA treatment and to 62.50% with 2000 ppm IBA treatment. In another work, it was revealed that the rooting percentage of 6.7% in the control group of plum cuttings increased up to 46.7% with IBA treatment [84].

Edizer and Demirel [85] reported that IBA treatments at 2000, 3000 and 4000 ppm significantly increased rooting percentage in cuttings of cherry, peach and two different plum species. They indicated that the rooting percentage of 15% in the control group of peach cuttings increased up to 80, 85 and 86.67% with IBA treatments at 2000, 3000 and 4000 ppm, respectively. In the same work, the rooting percentages of 46.67 and 60% in the control groups of plum cuttings and the rooting percentage of 48.34% in the control group of cherry cuttings were found to be 90% with 3000 ppm IBA treatment [85]. Similarly, there are studies reporting that IBA significantly increased rooting percentage in sage cuttings, and the rooting percentage of 16.25% in the control groups was increased up to 78.75% with 100 ppm IBA treatment [86].

Naphthalene acetic acid, which has a significant effect on rooting, was used in this report. It is a synthetic hormone that has been used in orcharding for many years for thinning heavy fruit set. With the effect of thinning, fruit size and quality can be increased [61]. However, the positive effect of NAA on rooting was also proven by many reports. In a study, the rooting percentage of 51.14% in the control group of *Schefflera arboricola* L. was found to have increased to 75% with 5000 ppm NAA treatment [87]. Similarly, a 6-h NAA treatment at 250 ppm was reported to increase the rooting percentage from 0 (control group) to 23% in cuttings of *Capparis sipinosa* L. [88].

Among the natural plant-growing regulators, gibberellins are the third most widely used class (estimated rate of 17%) [65], especially for horticultural applications. GA3 is mostly produced by fermentation from the fungus *Gibberella*. Today, there are approximately 100 known GA molecules, more than 50 found in plant seeds. GA3 is the most widely used one for commercial purposes [64].

There are many reports examining the effect of GA3 on rooting such as the one by Hepaksoy [89] on *Prunus avium* L. and *Prunus mahaleb* L., the work by Aygün and Dumanoğlu [90] on *Cydonia oblonga* Miller., the research by Coşge et al. [91] on *Capparis ovata* Desf. and *Capparis ovata* L., the work by Selby et al. [92] on *Picea sitchensis* (Bong.) Carr, the study by Sevik et al. [87] on *S. arboricola* L. and the research by Sevik and Guney [93] on *Melissa officinalis* L. Although it has not been determined that GA3 has a significant effect in many species, there are studies showing that it increases rooting percentage to a considerable extent [61, 94].

This work found that hormone treatments affected root number to a considerable extent. The root number, which was 1.3 in the control group, increased to 1.5 and 1.7 with 5000 ppm IAA treatment and 5000 ppm NAA treatment, respectively.

The findings of this report are in conformity with other works in the literature. Sevik and Turhan [95] reported that all hormone treatments increased the number of roots, and they found that the average root number of 0.97 in the control group of *L. martagon* was increased to 1.33 with 3000 ppm IBA + 1000 ppm NAA treatment. They also reported that the average root number of 1.14 in the control group of *L. artvinense* was increased to 2.50 with 1000 ppm IBA treatment [76]. Guney et al. [61] also reported that the root number of 1.2 in the *L. martagon* seeds increased up to 2.0 with 3000 ppm GA3 treatment.

The reports conducted on other plant species also show that root number can be increased with hormone treatments. Sevik and Guney [93] reported that the root number of 2.67 in the control group of *M. officinalis* L. was increased to 5.5 with 5000 ppm IBA treatment. Yildiz [96] also reported that the root number of 1.67 in the control group of plum was increased to the average number of 2.3 in the seeds treated with IBA, whereas Ayanoğlu and Özkan [86] indicated that the root number of 4.22 in the control group of *Salvia officinalis* L. was increased to 22.35 with 100 ppm IBA treatment. Similarly, there are studies revealing that IBA treatment can increase the number of roots four to five times in peach, plum and cherry cuttings [85]. In the work by Demiral and Ülger [97], the root number of 4.40 in the control group of cherry cuttings was increased to 16.29 with 6 mg/l NAA treatment, while in the research by Sevik et al. [87] the average root number of 5.82 in the control group of *S. arboricola* L. was increased to 9.63 and 12.33 with 1000 ppm NAA and GA3 treatments, respectively. In the report by Polat et al. [83], the root number of 0.38 in the control group of plum cuttings was increased to 10.43 with 2000 ppm IBA treatment, while in the research by Sevik and Guney [98], the root number of 4.8 in the control group of *M. officinalis* L. could be increased to 12.5 with 1000 ppm IAA treatment.

The findings of this study also reveal that hormone treatments have affected the root length of the plants to a considerable extent, increasing the root length of 27.153 mm in the control group up to 66.419 mm with 5000 ppm IBA treatment. However, more importantly, root lengths were found to be higher than those in the control groups after all hormone treatments. Therefore, the treatments can be said to have a positive effect on the root length.

The highest root lengths were obtained with 3000 ppm IBA treatment in the work by Turhan [65] on *L. martagon*, in the report by Guney et al. [61] on *L. martagon* seeds and in the research by Topacoglu et al. [99] on *Ficus benjamina* cuttings.

The works conducted show that the root length can be increased to 2.75 times in *L. artvinense* with 3000 ppm GA3 treatment [76], 16.17 times in plum cuttings with 2000 ppm IBA treatment [83], 2.34 times in cherry cuttings with 4000 ppm IBA treatment [85], 1.63 times in *S. arboricola* with 1000 ppm IBA treatment [10] and 2.26 times in sage with 100 ppm IBA [87], as compared to the control group.

Besides, results also show that, just like the root length, the highest stem height and diameter values have been obtained with 5000 ppm IBA treatment. Stem height and diameter values are 34 and 27% higher than those of the control group, respectively. Therefore, the treatments can be said not to have such a huge effect on stem formation as to make a significant difference. Similar results were also obtained from *L. martagon* seeds. The stem height of 7.16 mm in the control group of *L. martagon* increased to 8.17 mm with 3000 ppm IBA treatment, and the stem diameter of 2.93 mm increased to 3.78 with 5000 ppm IBA treatment [61]. Likewise, Turhan [65] reported that the stem number of 0.82 in the control group of *L. martagon* increased to 0.91 with 1000 ppm IBA + 1000 ppm IAA treatment, the stem height of 9.05 mm increased to 9.19 mm with 1000 ppm IBA + 1000 ppm NAA treatment and the stem diameter of 2.40 mm increased to 2.81 mm with 3000 ppm IBA + 1000 ppm NAA treatment.

However, it was indicated that, in *L. artvinense* plants treated with hormones, the stem number of 0.43 was increased to 0.92, the stem height of 1.53 mm was increased to 6.55 mm and the stem diameter of 0.97 mm was increased to 4.3 mm [76].

Apart from these, several other studies have been carried out on *Lilium* cultivation; however, most of them focus on microculture techniques. There are also works conducted on *L. davidii* var. *unicolor* [100] and *L. longiflorum* with IBA and NAA treatment [101], on *L. davidii* var. *unicolor* [102], *L. oritential* and *L. longiflorum* with IAA, IBA and NAA treatment [103, 104], on *L. longiflorum* with IAA and IBA treatment [105] and on *L. japonicum* with BA and GA3 treatment [106]. However, it is difficult to compare results of these reports with those obtained in the present work, due to rather different experimental approach.

2. Conclusions

Collecting of endangered *Lilium* species for commercial purposes is a great harm to the natural populations. The most effective way to prevent the collection of endangered species from the wild is to establish simple, cost-effective and reproducible methods of cultivation. To date, most studies focus on *Lilium* cultivation using microculture techniques, which are costly and hard to apply for the villagers who collect the flowers from the wild. Therefore, the question is still open.

This report attempted to define simple, inexpensive and effective cultivation methods for *Lilium* species. The work found that hormones applied using simple mechanisms could increase rooting process to a great extent, as well as enhancing the morphological traits of newly generated plants.

The results of this study show that all hormones have been extremely effective especially in increasing germination percentage. The germination percentage of 40% in the control group increased up to 60.66, 75.34, 82.66 and 84% in the seeds treated with NAA, IAA, IBA and GA3, respectively. Besides, 100% germination percentage was achieved with 5000 ppm GA3 treatment.

In addition, the results also reveal that hormone treatments affect each treatment differently. For example, the highest rooting percentage was obtained with 5000 ppm GA3 treatment. On the other hand, the highest number of roots was obtained with 5000 ppm NAA treatment, while the highest root length, stem height and stem diameter values were obtained with 5000 ppm IBA treatment. In practice, it would be best to use the hormone which has the greatest effect on the treatment that is desired to be enhanced.

Similar works may be repeated especially on species that are important in terms of mass cultivation, such as ornamental plants or medicinal or aromatic plants, and important data may be acquired as a result. Thus, further studies will help innovating effort, time and cost-saving practices in plant cultivation.

Author details

Kerim Guney[1], Mehmet Cetin[2*], Hakan Sevik[3] and Kudret Betül Güney[4]

*Address all correspondence to: mehmet.cetin@temple.edu

1 Faculty of Forestry, Department of Forestry, Kastamonu University, Kastamonu, Turkey

2 Faculty of Engineering and Architecture, Department of Landscape Architecture, Kastamonu University, Kastamonu, Turkey

3 Faculty of Engineering and Architecture, Department of Environmental Engineering, Kastamonu University, Kastamonu, Turkey

4 Graduate School of Natural and Applied Science, Ankara University, Ankara, Turkey

References

[1] Yüksel ÜD: Investigation of air temperature and surface temperatures in built-up and green areas in the cities: the case of Ankara city. Ecology. 2008; 18 (69): 66–74 (in Turkish).

[2] Sevik H, Karakaş H, Şenöz E: Evaluation of air quality in terms of the amount of carbon dioxide in black sea region. International Journal of Engineering Science & Research Technology. 2013; 2 (2): 173–177.

[3] Konijnendijk CC: A decade of urban forestry in Europe. Forestry Policy and Economics. 2003; 5 (2): 173–186.

[4] Yücel E, Ocak A, Özkan K, Soydam S: In Turkey as an ornamental plant grown trees and shrubs, III. Ornamentals National Congress, Ege University, Faculty of Agriculture, Department of Horticulture, Izmir, 8–10 November 2006.

[5] Öztürk S, Bozdoğan E: Determination of the perceived quality of urban life in new and traditional housing textures. Fresenius Environmental Bulletin. 2014; 23 (10): 2415–2421.

[6] Papinchak H, Holcomb EJ, Orendovici BT, Decoteau, DR: Effectiveness of houseplants in reducing the indoor air pollutant ozone. HortTechnology. 2009; 19 (2): 286–290.

[7] Tani A, Hewitt CN: Uptake of aldehydes & ketones at typical indoor concentrations by houseplants. Environmental Science and Technology. 2009; 43 (21): 8338–8343.

[8] Cetin M: Chapter 55: Using Recycling Materials for Sustainable Landscape Planning. Environment and Ecology at the Beginning of 21st Century. Ed.: Recep Efe, Carmen

Bizzarri, İsa Cürebal, Gulnara N. Nyusupova, ST. Kliment Ohridski University Press, Sofia; 2015. p. 783–788, ISBN: 978-954-07-3999-1.

[9] Öztürk S, Bozdoğan E: The contribution of urban road trees on improving the air quality in an urban area. Fresenius Environmental Bulletin. 2015; 24 (5a): 1822–1829.

[10] Sevik H, Cetin M, Belkayali N: Effects of forests on amounts of CO_2: case study of Kastamonu and Ilgaz Mountain National Parks. Polish Journal of Environmental Studies. 2015; 24 (1): 253–256.

[11] Yiğit N, Öztürk A, Sevik H: Ecological impact of urban forests (example of Kastamonu urban forest). International Journal of Engineering Sciences & Research Technology. 2014; 3 (12): 558–562.

[12] Sevik H, Belkayali N, Aktar G: Change of chlorophyll amount in some landscape plants. Journal of Biotechnological Sciences. 2014; 2 (1): 10–16.

[13] Belkayali N, Kesimoglu MD: The stakeholders' point of view about the impact of recreational and tourism activities on natural protected area: a case study from Kure Mountains National Park, Turkey. Biotechnology & Biotechnological Equipment. 2015; 29 (6): 1092–1103.

[14] Cetin M: Evaluation of the sustainable tourism potential of a protected area for landscape planning: a case study of the ancient city of Pompeiopolis in Kastamonu. International Journal of Sustainable Development & World Ecology. 2015; 22 (6): 490–495. DOI: 10.1080/13504509.2015.1081651

[15] Cetin M: Determining the bioclimatic comfort in Kastamonu City. Environmental Monitoring and Assessment. 2015; 187 (10): 640. DOI: 10.1007/s10661-015-4861-3

[16] Turna İ, Güney D: Altitudinal variation of some morphological characters of scots pine (*Pinus sylvestris* L.) in Turkey. African Journal of Biotechnology. 2009; 8 (2): 202–208.

[17] Tekçe N, Belkayali N, Oğuz D, Baştemur CT: A survey on recreational use of domestic water supply reservoirs: a case study from Kurtboğazi–Ankara, Turkey African Journal of Agricultural Research. 2010; 5 (14): 1897–1907.

[18] Talay İ, Kaya F, Belkayalı N: Socio-economic structure effects on the recreational trends and demands: a case study Bartin city. Journal of Geographical Sciences. 2010: 8 (2): 147–156 (in Turkish).

[19] Djukanovic R, Wargocki P, Fanger PO: Cost-benefit analysis of improved air quality in an office building. Proceedings of Indoor Air. 2002; 1: 808–13.

[20] Lohr VI, Pearson-Mims CH, Goodwin GK: Interior plants may improve worker productivity and reduce stress in a windowless environment. Journal of Environmental Horticultural. 1996; 14: 97–100.

[21] Chang C, Chen P: Human response to window views and indoor plants in the workplace. HortScience. 2005; 40: 1354–1359.

[22] Lohr VI, Pearson-Mims CH, Goodwin GK: Interior plants may improve worker productivity and reduce stress in a windowless environment. Journal of Environmental Horticulture. 1996; 14: 97–100.

[23] Ay S: The export, problems and solution proposals of ornamental plants: investigation of sampling in Yalova. Suleyman Demirel University. The Journal of Faculty of Economics and Administrative Sciences. 2009; 14 (3): 423–443.

[24] Öztürk S: Determining management strategies for the Sarikum Nature Protection Area. Environmental Monitoring and Assessment. 2015; 187 (3); 113.

[25] Šálek L, Güney K: New host plant for the species *Agapanthia lateralis Gangl*. (Coleoptera; Cerambycidae). Entomological News. 2014; 124 (1): 29–32.

[26] Oğuz D, Diriöz ED, Belkayalı N: Tourists' perception of landscape design: the case of resorts in the Belek specially protected area. African Journal of Agricultural Research. 2010; 5 (10): 1028–1035.

[27] Belkayali N, Sevik H, Guloglu Y: What affects perceptions of local residents toward protected areas? A case study from Kure Mountains National Park, Turkey. International Journal of Sustainable Development & World Ecology. 2016; 23 (2): 194–202. DOI: 10.1080/13504509.2015.1098743

[28] Aricak B, Kucuk O, Enez K: Determining a fire potential map based on stand age, stand closure and tree species, using satellite imagery (Kastamonu central forest directorate sample). Croatian Journal of Forest Engineering. 2014; 35 (1): 101–108.

[29] Canbulat S: Checklist of Turkish Raphidioptera on the basis of distribution pattern and biogeographical analysis. Turkish Journal of Zoology. 2015; 39(2): 225–234.

[30] Cetin M, Sevik H: Evaluating the recreation potential of Ilgaz Mountain National Park in Turkey. Environmental Monitoring and Assessment. 2016; 188 (1): 52. DOI: 10.1007/s10661-015-5064-7

[31] Gumus S, Aricak B, Enez K, Acar HH: Analysis of tree damage caused by rockfall at forest road construction works. Croatian Journal of Forest Engineering. 2009; 30 (2): 151–158.

[32] Aricak B: Using remote sensing data to predict road fill areas and areas affected by fill erosion with planned forest road construction: a case study in Kastamonu Regional Forest Directorate (Turkey). Environmental Monitoring and Assessment. 2015; 187 (7): 1–10.

[33] Sevik H, Cetin M: Effects of water stress on seed germination for select landscape plants. Polish Journal of Environment Studies. 2015; 24 (2): 689–693.

[34] Topacoglu O, Sevik H, Akkuzu E: Effects of water stress on germination of Pinus nigra Arnold. Seeds. Pakistan Journal of Botany. 2016; 48(2): 447–53.

[35] Kesik HI, Korkut S, Hiziroglu S, Sevik H: An evaluation of properties of four heat treated wood species. Industrial Crops and Products. 2014; 60: 60–65.

[36] Öztürk S, Bilgili B: The use of natural plant species of Kastamonu University campus in landscape. Kastamonu University Journal of Forestry Faculty. 2015; 15 (1): 175–182.

[37] Sevik H, Topacoglu O: Variation and ınheritance pattern in cone and seed traits of Scots pine (*Pinus sylvestris* L.) for evaluation of genetic diversity. Journal of Environmental Biology. 2015; 36 (5): 1125–1130.

[38] Sevik H, Ayan S, Turna İ, Yahyaoğlu Z: Genetic diversity among populations in Scotch pine (*Pinus silvestris* L.) seed stands of Western Black Sea Region in Turkey. African Journal of Biotechnology. 2010; 9 (43): 7266–7272.

[39] Guney D, Yahyaoglu Z, Turna I, Muller-Starck G: Genetic variation in *Pinus brutia* in Turkey. Fresenius Environmental Bulletin. 2014; 23 (5): 1249–1254.

[40] Güney D, Hatipoğlu E, Atar F, Turna İ, Kulaç, Ş: Changes of some morphological traits of *Oriental Hornbeam* (*Carpinus orientalis* Miller) seeds depending on altitude. In *International Caucasia Forestry Symposium*, 2013, July.

[41] Kaya LG, Cetin M, Doygun H: A holistic approach in analyzing the landscape potential: Porsuk Dam Lake and its environs, Turkey, Fresenius Environmental Bulletin. 2009; 18 (8): 1525–1533.

[42] Cetin M: Chapter 27: Landscape Engineering, Protecting Soil, and Runoff Storm Water. Advances in Landscape Architecture, InTech, Ed: Murat Özyavuz, 2013, p. 697–723. DOI: 10.5772/55812

[43] Cetin M: Using GIS analysis to assess urban green space in terms of accessibility: case study in Kutahya. International Journal of Sustainable Development & World Ecology. 2015; 22 (5): 420–424. DOI: 10.1080/13504509.2015.1061066

[44] Enez K, Aricak, B, Sariyildiz T: Effects of harvesting activities on litter decomposition rates of scots pine, trojan fir, and sweet chestnut. Šumarski List. 2015; 139 (7–8): 361–367.

[45] Guloglu Y: The establishment of forest ownership and the legal regulations on the forests until the Tanzimat (Reform) period in the Ottoman State. Kastamonu University. Journal of Forestry Faculty. 2015; 10 (2): 180–194.

[46] Guloglu Y: Evaluation of forest land allocation for tourism with regard to agents of administrative act. Kastamonu University Journal of Economics and Administrative Sciences. 2015; 10: 45–55.

[47] Ozturk S, Enez K: Determination of the perceptions of secondary education students towards environment and nature. Journal of Environmental Protection and Ecology. 2015; 16 (2): 723–732.

[48] Olmez Z, Gokturk A, Temel F: Effects of cold stratification, sulphuric acid, submersion in hot and tap water pretreatments on germination of bladder-senna (Colutea armena Boiss. & Huet.) seeds. Seed Science and Technology. 2007; 35 (2): 266–271.

[49] Olmez Z, Gokturk A, Temel F: Effects of some pretreatments on seed germination of nine different drought-tolerant shrubs. Seed Science and Technology. 2007; 35 (1): 75–87.

[50] Güney D, Kulaç Ş, Turna İ: Study on cone formation stage of Caucasian Fir (*Abies nordmanniana* ssp. nordmanniana). Kastamonu University Journal of Forestry Faculty. 2012; 12 (3): 228–233.

[51] Gokturk A, Olmez Z, Karasah B, Surat H: Effects of cold stratification and sulphuric acid pre-treatments on germination of pomegranate (*Punica granatum* L.) seeds in greenhouse and laboratory conditions. Scientific Research and Essays. 2012; 7 (25): 2225–2229.

[52] Olmez Z, Gokturk A, Gulcu S: Effects of cold stratification on germination rate and percentage of caper (*Capparis ovata* Desf.) seeds. Journal of Environmental Biology. 2006; 27 (4): 667–670.

[53] Kulaç Ş, Nzokou P, Guney D, Cregg BM, Turna I: Growth and physiological response of fraser Fir [*Abies fraseri* (Pursh) Poir.] seedlings to water stress: seasonal and diurnal variations in photosynthetic pigments and carbohydrate concentration. HortScience. 2015; 47 (10): 1512–1519.

[54] Kulac S, Turna I, Çiçek E, Sağlam A, Taşdemir, Ü: Arly field performance of drought-stressed scots pine (*Pinus sylvestris* l.) seedlings. Pakistan Journal of Botany. 2015; 47 (4): 1317–1324.

[55] Boydak M, Dirik H, Tilki F, Çalıkoğlu M: Effects of water stress on germination in six provenances of *Pinus brutia* seeds from different bioclimatic zones in Turkey. Turkish Journal of Agriculture and Forestry. 2003; 27 (2): 91–97.

[56] Ölmez Z, Akın E, Göktürk A: Effects of different growing media and polyethylene pot size on some morphological seedling characteristics of Caper (*Capparis ovata* Desf.). Artvin Coruh University Journal of Forestry Faculty. 2011; 12 (2): 101.

[57] Gao YD, Harris AJ, Zhou SD, He XJ: Evolutionary events in *Lilium* (including Nomocharis, Liliaceae) are temporally correlated with orogenies of the Q–T plateau and the Hengduan Mountains. Molecular Phylogenetics and Evolution. 2013; 68: 443–460.

[58] Godo T, Kobayashi, K, Tagami T, Matsui K, Kida T: In vitro propagation utilizing suspension cultures of meristematic nodular cell clumps and chromosome stability of Lilium=formolongi hort. Scientia Horticulturae. 1998; 72: 193–202.

[59] Lian ML, Chakrabarty D, Paek KY: Growth of *Lilium* oriental hybrid 'Casablanca' bulblet using bioreactor culture. Scientia Horticulturae. 2003; 97: 41–48.

[60] Kumar S, Kanwar J, Sharma D. In vitro propagation of *Lilium*. Advances in Horticultural Science. 2006; 20: 181–188.

[61] Guney K, Cetin M, Sevik H, Guney KB: Influence of germination percentage and morphological properties of some hormones practice on Lilum martagon L. Seeds. Oxidation Communications. 2016 Jan 1; 39(1): 466–74.

[62] The Plant list. A working list of all plant species. The plant list- *Lilium artvinense* Miscz. [Internet]. 2016. Available from: http://www.theplantlist.org/tpl1.1/record/kew-280712 [Accessed: 2016-01-15].

[63] Mitchell A: Thoughts on the classification of species lilies past and present, Sectional Species List. Lilies and Related Plants. Ed.: A Mitchell. The Royal Horticultural Society LILIY Group, 2009; p. 113–119.

[64] Kumlay AM, Eryiğit T: Growth and development regulators in plants: plant hormones. Iğdır University Journal of Research Institute of Science and Technology. 2011; 1 (2): 47–56 (in Turkish).

[65] Turhan H: Effects of IBA (indole butyric acid) on rooting and newly stem to *Turkish lili (Lilium martagon* L.) onion [thesis]. Graduate School of Natural and Applied Sciences Department of Forest Engineering: Kastamonu University; Msc Thesis, 201552 p. (in Turkish).

[66] Seyedi A, Esmaeili A, Zadeh K, Posiabidi M: Comparative evaluation of the rooting in cuttings in (*Bougainvillea glabra* L.). International Journal of Farming and Allied Sciences. 2014; 3 (8): 872–875.

[67] Klein JD, Cohen S, Hebbe Y: Seasonal variation in rooting ability of Myrtle (*Myrtus communis* L.) cuttings. Scientia Horticulturae. 2000; 83: 71–76.

[68] Sağlam AC, Yaver S, Başer İ, Cinkılıç L: The effects of different hormones and their doses on rooting of stem cuttings in Anatolian sage (Salvia Fruticosa Mill.). APCBEE Procedia. 2014; 8: 348–353.

[69] Miransari M, Smith DL: Plant hormones and seed germination. Environmental and Experimental Botany. 2014; 99: 110–121.

[70] Graeber K, Nakabayashi K, Miatton E, Leubner-Metzger G, Soppe W: Molecular mechanisms of seed dormancy. Plant Cell and Environment. 2012; 35: 1769–1786.

[71] Santner A, Calderon-Villalobos L, Estelle M: Plant hormones are versatile chemical regulators of plant growth. Nature Chemical Biology. 2009; 5: 301–307.

[72] Went FW: On a substance causing root formation. Proceedings of the Section of Sciences, Koninklijke (Nederlandse) Akademie van Wetenschappen, Amsterdam. 1929; 32: 35–39.

[73] Grunewald W, Noorden GV, Isterdael GV, Beeckman T, Gheysen G, Mathesius U. (2009, September): Manipulation of auxin transport in plant roots during Rhizobium symbiosis and nematode parasitism. The Plant Cell. 2009; 21: 2553–2562.

[74] Cetin V: Plant growth regulators in used fruits and vegetables. Food and Feed Science-Technology: 2002; 2: 40–50.

[75] Şeker M, Akçal A, Sakaldaş M, Gündoğdu M A: Effects of different periods and auxins concentrations on rooting ability of strawberry tree (*Arbutus unedo* L.) cuttings. Journal of Agricultural Faculty of Uludag University. 2010; 24 (1): 99–108.

[76] Sevik H, Cetin M: Effects of some hormone applications on germination and morphological characters of endangered plant species Lilium artvinense I. onion scales. Bulgarian Chemical Communications. 2016; 48(2): 259–263

[77] Cerveny C, Gibson J: Rooting hormones growers 101. Crop Cultivation. 2005; 1:36–44.

[78] Swamy SL, Puri S, Singh AK: Effect of auxins (IBA and NAA) and season on rooting of juvenile and mature hardwood cuttings of *Robinia pseudoacacia* and *Grewia optiva*. New Forests. 2002; 23 (2): 143–157.

[79] Stefancic M, Stampar F, Osterc G: Influence of IAA and IBA on root development and quality of Prunus "GiSelA 5" leafy cuttings. HortScience. 2005; 40 (7): 2052–2055.

[80] Chhun T, Taketa S, Tsurumi S, Ichii M: The effects of auxin on lateral root initiation and root gravitropism in a lateral rootless mutant Lrt1 of rice (*Oryza sativa* L.). Plant Growth Regulation. 2003; 39: 161–170.

[81] Nordström AC, Jacobs FA, Eliasson L: Effect of exogenous indole-3-acetic acid and indole-3-butyric acid on internal levels of the respective auxins and their conjugation with aspartic acid during adventitious root formation in pea cuttings. Plant Physiology. 1991; 96: 856–861.

[82] Zenginbal H, Haznedar A: An investigation on the propagation of tea (*Camellia sinensis* L.) by grafting and budding methods. Academic Journal of Agriculture. 2013; 2 (2): 99–106 (in Turkish).

[83] Polat AA, Kamiloğlu Ö, Durgaç C: The effects of indole butyric acid (IBA) on rooting of wood cuttings of can plums in open air. MKU Journal of Agricultural Faculty. 1997; 2 (2): 103–114 (in Turkish).

[84] Yıldız K: The effect of IBA, CEPA and AVG on rooting of Hardwood cuttings in different fruit species. Journal of Agricultural Science 2001; 11 (1): 51–54.

[85] Edizer Y, Demirel MA: A study on the some characteristics of rooting of green cuttings of the some clonal rootstock ın mist propagation. GOU Journal of Agriculture. 2012; 29 (2): 1–8 (in Turkish).

[86] Ayanoğlu F, Özkan FC: Change in tissue mineral elemental concentration during root initiation and development of *Salvia officinalis* L. cuttings and IBA effects. Turkish Journal of Agriculture and Forestry. 2000; 24: 677–682 (in Turkish).

[87] Sevik H, Güney K, Topaçoğlu O, Ünal C: The influences of rooting media and hormone applications on rooting percentage and some root characters in *Schefflera arboricola*. International Journal of Pharmaceutical Science Invention. 2015; 4 (2): 25–29.

[88] Söyler D, Arslan N: The effects of some plant growing regulators on the rooting of capers (*Capparis spinosa* L). Turkish Journal of Agriculture and Forestry. 2000; 24: 595–600 (in Turkish).

[89] Hepaksoy S: Investigations on micropropagation of some cherry rootstocks I. shoot development and proliferation. Ege University Journal of Agriculture Faculty. 2004; 41 (3): 11–22 (in Turkish).

[90] Aygün A, Dumanoğlu H: Shoot organogenesis from leaf discs in some quince (*Cydonia oblonga* Mill.) genotypes. Journal of Agriculture Sciences, 2006; 13 (1): 54–61 (in Turkish).

[91] Coşge B, Gürbüz B, Söyler D, Şekeroğlu N: Caper (*Capparis* spp.) cultivation and its importance (a review). Journal of Crop Research. 2005; 2: 29–35.

[92] Selby C, Kennedy J, Harvey MR: Adventitious root formation in hypocotyl cuttings of *Picea Sitchensis* (Bong.) Carr. —the influence of plant growth regulators. New Phytologist. 1992; 120 (4): 453–457.

[93] Sevik H, Guney K: Effects of IAA, IBA, NAA, and GA3 on rooting and morphological Features of *Melissa officinalis* L. stem cuttings. The Scientific World Journal. 2013; 2013: 5. http://dx.doi.org/10.1155/2013/909507

[94] Ünal O, Gökceoğlu M, Topcuoğlu ŞF: An investigation on seed germinating and reproduction with cutting of endemic origanum species in Antalya. Akdeniz University Journal of Agriculture Faculty. 2004; 17 (2): 135–147 (in Turkish).

[95] Sevik H, Turhan H: Effects of IBA (indole butyric acid) on rooting and newly stem to Turkish Lili (*Lilium martagon* L.) onion. The International Conference on Science, Ecology and Technology I (Iconsete'2015 – Vienna) Abstract Book, p. 61, August 25–28, 2015, Vienna, Austria.

[96] Yildiz K: The effect of IBA, CEPA and AVG on rooting of hardwood cuttings in different fruit species. Yuzuncuyil University Journal of Agricultural Sciences. 2001; 11 (1): 51–54.

[97] Demiral S, Ülger S. A research on propagation of Gisela 5 cherry rootstock by tissue culture. Akdeniz University Journal of Agriculture Faculty. 2008; 21 (1): 117–121 (in Turkish).

[98] Sevik H, Guney K: Effects of some hormone applications on morphological features of *Melissa officinalis* L. root cuttings. Soil-Water Journal. 2013, 2(2): 1647–1652.

[99] Topacoglu O, Sevik H, Guney K, Unal C, Akkuzu E, Sivacioglu A: Effect of rooting hormones on the rooting capability of Ficus benjamina L. cuttings. Šumarski list. 2016 Feb 29; 140(1–2): 39–44.

[100] LingFei X, FengWang M, Dong L: Plant regeneration from in vitro cultured leaves of *Lanzhou lily* (*Lilium davidii* var. unicolor). Scientia Horticulturae. 2009; 119: 458–461.

[101] Bacchetta L, Remotti P, Bernardini C, Saccardo F: Adventitious shoot regeneration from leaf explants and stem nodes of *Lilium*. Plant Cell, Tissue and Organ Culture. 2003; 74 (1): 37–44.

[102] Jing L: Optimization of highly efficient plant regeneration system for *Lilium davidii* var. unicolor (Hoog) cotton. Journal of Southern Agriculture. 2011; 42 (8): 839–842.

[103] Chen L-J, Zhang X-G, Ma S, Zhong M, Guo Z-F, Ming J: Study on the rapid propagation of *Lilium oritential* and *Lilium Longiflorum*. Northern Horticulture. 2010; 14: 682. DOI: 10.11937/bfyy.2010014037

[104] Yan FY, Hu XY, Pei XH, Yin DS: Effect of hormone on scale propagation of Lilium oriental [J]. Liaoning Agricultural Sciences. 2008; 6: 1–5.

[105] Han Z: Fossilization: five central issues. International Journal of Applied Linguistics. 2004; 14 (2): 212–242. DOI: 10.1111/j.1473-4192.2004.00060.x

[106] Yamagishi M: Effects of cold treatment, BA and GA3 on enlargement and leaf emergence of in vitro cultured bulblets of *Lilium japonicum* Thunb. Journal of the Japanese Society for Horticultural Science. 1995; 64 (2): 367–373.

Chapter 5

Postharvesting Techniques and Maintenance of Seed Quality

Zubaida Yousaf, Nadia Saleh, Asba Ramazan and Arusa Aftab

Additional information is available at the end of the chapter

http://dx.doi.org/10.5772/64994

Abstract

Healthy seeds and grains are the demanding enterprise of the recent era for the production of high yield in the next season. The seeds must be stored for the maintenance of high-yielding crop. During storage, major losses of seeds are caused by various biological and nonbiological factors. There is a need to examine reasonable factors of these crop losses, which ultimately affect the market value and quality of the seed. The quality of seeds can be maintained by using careful postharvest handling techniques. There is need to establish the well-suited methods to assess the losses during the process and to use the best technique to minimize the loss and to ensure the quality and safety of the crop. The target is to achieve the high-quality seeds of the national and international standards that could meet the demand of the supplier. This chapter emphasizes on the aspects and postharvest techniques that are used to maintain seed quality. A comprehensive review of the better, economical, convenient, and productive methods is provided, focused on the needs of developing countries but also with relevance in more industrialized countries.

Keywords: crop maintenance, handling techniques, storage losses, postharvest

1. Introduction

Seed industry survives for the sake of better quality of seeds and postharvest storage techniques. It is a seed quality program that seeks the high achievement and gives the quality assurance. Seed quality is one of the most demanding enterprises of present days. The majority of the seeds especially cereals and legumes by small-scale farmers are produced and stored on the farms.

The major problem is the damage due to biological factors such as molds, and insects can be alleviated by implicating effective storage techniques [1–4]. It is the foremost priority of the farmers to prevent the crop losses and storage of seeds/grains. Farmers often purchase the new seeds/grains from the market to produce the next crop for higher yield due to the risk of crop loss. There is a need to develop the effective strategies for seed storage that can give the confirmation of better crop yield and reduce the risks of storage losses [5].

1.1. Valuable seeds: achieving seed quality

The value of the seeds allows the farmers to produce high-yielding crops from good quality healthy seeds. Seed quality is the measure of the potential to produce desired quality, healthy, and high-yielding crops at low planting rates [5]. Seed quality can neither be obtained automatically nor by a permanent process. There is nature's pressure against the quality of the seeds. Efforts are being made to produce best quality seeds for the farmers. Any handling or production stage can reduce the quality of the seed. To alleviate these losses, careful technical management is required for all seed operations [6].

1.2. Principles of seed storage

Seeds when stored in natural environment or ambient temperature respond readily to temperature, available oxygen, and relative humidity. Metabolic activities, age, and longevity of seeds can be manipulating by controlling the humidity temperature and oxygen [7]. Reduction of seed moisture content up to an appropriate limit is prerequisite for storage as seed could be damaged because of desiccation. Seeds can be stored for a longer period due to lower level of humidity. According to the thumb rule, the life of the seed doubles by decreasing the moisture content to 1% in case if seed moisture content is between 5 and 14%. Higher moisture content is more affected by higher temperature so seeds need to be stored in cool location. The life of the seed doubles by decreasing the temperature to 5°C and is applicable between 0 and 50°C [8].

Oxygen level can be controlled by hermetic storage in a sealed container which reduces the physiological aging of the grains as well as reduces the physical damage due to insects and fungal growth [8].

2. Postharvest techniques of seed storage

2.1. Drying

Physiological maturity attained by the cereal and legumes at moisture content between 35 and 45% is crop dependent. Temperature affects the storage of seeds at moisture content between 10 and 14%. For high-quality yield of crops, timely harvesting and drying are necessary. Biologically active seeds deteriorate readily under most of the circumstances due to fungal contamination and attack of insects and other pests. The main purpose of drying is to reduce the respiration in seeds [6, 9]. The process also impedes qualitative damage due to fungi and other insect pests. Drying can itself affect the quality of the seeds. Extensive drying under very high temperature can damage the seed. In summer season, simple drying methods are used through exposure to sun and adequate wind. The alternative drying methods have been devised for high-yielding varieties and improved farming practices and irrigation to deal with increased production or harvesting in wet season in multi-cropping [10, 11].

2.2. Sun drying

Sun drying of seeds is the drying method in the tropical developing countries. The method is employed when crop is ready for the harvest. Some seeds like maize can be sun dried although during drying, crops get sensitive to insect infestation, rodent or birds' attack, and mold damage. A threshed seed drying by spreading on sheets or tray is a common phenomenon but has a risk of soil or stone contamination. For example, paddy at rice mills is at large quantity [6]. Rice is dried on especially build drying floor that allow easy run off of rain water. The seeds are dispersed in thin layer and turned at regular intervals to facilitate drying and covered at night with the help of sheets. There are some disadvantages of the process that temperature is an uncontrollable factor. In paddy rice, high temperature can cause stress or cracks in the seeds which lead to high level of damage during milling. Yield can be contaminated by dust, atmospheric contamination or insect infestation, or human or animal disturbance [6, 9].

2.3. Solar drying

It is the modification of sun drying in which sun rays are collected in an especially designed unit for air removal in adequate ventilation system. The unit has 20–30° higher temperature than open drying, and less time is consumed in the process [12]. In solar dryers, solar collector is used to heat air which then allowed to pass to the seed beds. It comprises two basic designs: natural convection dryers use thermal gradients and forced convection dryers force the air through solar collectors and seed layers. These dryers are suitable for farm use. The former design of the Asian Institute of Technology in Bangkok has been used as blueprint for several convection dryers and comprises a drying bin, a solar chimney, and a solar collector [13]. The solar collector consists of black polythene sheet or layer of burnt paddy husk; it is covered with clear polythene sheet. Perforated platform presents in drying bin. The disadvantages of the process are as follows: high structural profile, stability problem in windy condition, and the need of replacement of polythene sheets at regular intervals [6].

2.3.1. Mechanical dryers

The same principle of drying is used by mechanical dryers as forced convection solar dryer; the dryer forced the air through the seed bed and the air is heated with the help of a flat plate instead of conventional means. In modern automated storage system, drying takes place at one of the two points either in prestorage dryers (prior to loading seeds in freestanding loading) or in store dryer (after loading in final storage compartment) [9]. In prestorage dryers, ambient air is used in continuous flow dryers, and heat is generated by thermostatically controlled furnace which is powered by electricity, diesel, or gas. Heat may be supplied by direct or indirect way. Indirect way is preferred due to the separate outlet for the combustion product not through seeds. In the batch dryers, seeds are fed into properly defined batches, whereas in continuous flow dryers, the grains are flowed into the system and recovered at desired moisture content [6].

2.3.2. *Tray dryers*

Tray dryers are batch dryers of flat beds. The seeds are dispersed on the mesh tray at the depth of 600–700 mm, and warm dried air is passed through seeds to sufficiently dry them [6, 14].

2.4. Radial drying bin

Radial drying bin comprises two vertical metal mesh cylinders, one inside the other. Seeds are loaded between these two cylinders, and air is blown to the inner cylinder and passed from the inner to the outer mesh cylinder. By reversing air through seeds, air can be sucked from the central cylinder. However, there is the risk of overdrying of seeds in the inner cylinder which are in direct contact with the hot air. Air is wetter and cooler at the leaving side toward outside [6].

2.5. Continuous flow dryers

Moisture content of the seeds can be removed by sucking or blowing hot air by top-to-bottom passage through the system. Bin is present at the top of the drying section with cooling system at the base. Seed beds can be horizontal, vertical, or inclined. Seeds are moved by conveyors, scrapers, vibration, or gravity. The degree of drying depends upon the speed, size, and rate of flow of outlet conveyor of the dryer. Continuous flow dryer is varied by relative direction of air stream and seed flow [15]. Several continuous flow dry techniques are described below.

2.5.1. *Cross flow*

The seed passes through the two perforated sheets downward to the column by allowing horizontal passage of air through the seeds. The advantage of the dryer is that the moisture gradient can be defined at any stage of drying seeds [6].

2.5.2. *Counterflow*

A round bin is used to unload seeds with the upward flow of air. Little evaporative cooling takes place when hottest air passes through the driest seeds

2.5.3. *Concurrent flow*

In concurrent flow, wettest seeds are exposed to the hottest air during passage of air through seed bed. High temperature improves the efficiency of the dryer and cools the seeds by moisture evaporation [6, 16–18].

2.5.4. Cross flow

Cross-flow dryers have been used widely in recent years, but mixed-flow dryers have advantages over cross-flow dryers. Mixed flow dryers, combination of concurrent, counter and cross flow dryers has great advantage of efficient fuel consumption. But reduction in output because of uneven flow of seeds leading to uneven drying is one biggest hindrance in adaption of mixed flow drying [6].

2.5.5. Tower (mixed flow)

They consist of tall rectangular bins for storage, and triangular ducts are present along the width of the dryer at horizontal position. Half of the ducts are used for induction of warm air and removal of damped and cooled air is done by remaining ducts. It has multiple directions of seed flow and air flow [16–18].

2.5.6. Louvered bed dryer

The seeds are passed through different types of batches; the hot air is blown to the seeds to dry them. The dryers work on the principle of cross-flow dryer. The speech and depth of the drying beds depends upon the degree of drying. The two basic designs of the dryers are conveyor dryer and cascade dryer. The cascade dryers are gravity fed cross-flow dryers. The seed depth is controlled by roller dams and speed is controlled by output elevators. There can be incorporated changes to vary the length of the dryer. In conveyor dryer, air is blown to seeds through inclined louvered bed and the seed flow is controlled by variable speed, roller chain conveyor and heavy duty. These dryer can be one directional, two directional, or multidirectional; the variation in direction assists in removal of waste material and reduction in size of the dryer [6].

3. In store drying

This is an alternative method of drying in which seeds are load into bulks floor storage or in bin and then they are dried in stores [19].

3.1. Bulk on floor storage

They consist of especially strengthened wall which can bear the weight of seeds. The seeds are loaded in uniform depth. At one side of the building, fan is present for the aeration purpose and the plenum chamber runs along the center of the store or walls with perforated lateral ducts, below or above the floor level under the bulk of seeds [6].

3.2. In bin drying

This type of drying comprises one or more bins for drying purpose and other bins for storage. The dryers reduce the chance of physical damage due to the lesser handling. The shallow layer content of seed or insect may consume seed, leaving behind dust, weight loss cannot be detected easily [27]. To detect these losses, a useful mass of infested and noninfested seeds is ground into flour, and their weight is compared. A decrease in flour yield of infested mass than sound mass will be observed. Be aware of malpractices to compensate weight loss by adulteration of stones, earth, or sand and water. So there is a need to assess not only moisture changes but also the quantity of foreign matter present in the yield [23, 28].

4.5. Quality loss

Quality is an important trait considered by the consumers, and local traders have different criteria for assessment according to circumstances. Biochemical factors (acidity, sugars, flavor, and smell) involve size, shape, and appearance. Quality loss can also be due to contamination and foreign matter (insect fragments, rodent hair, excreta, weed seeds, earth, glass, stones, and parts of plants) content. Contaminants that are difficult to remove comprise soluble excretion of insects, pesticides, oils, toxins produced from fungal infections, and pathogenic organisms spread by rodents. Loss potential will be increased by raising the standard rules by consumers [24, 29].

4.6. Nutritional loss

The loss depends upon the qualitative and quantitative loss of nutritional value to the human population which affects the nutritional status of that population. This is mainly caused by feeding of pest on selective part of seed. *Plodia* and *Ephestia* feed selectively on the seed embryo and remove the vitamin and protein content. Many pests feed on bran of cereal seeds as a result reduced vitamin content. *Liposcelis* spp. selectively feeds on bran of rice and embryo (Pike, 1994). Weevil is an endosperm feeder and declines the carbohydrate content [6, 30].

4.7. Loss of seed viability

The loss of seed viability is linked with. The reasons of the losses may be temperature, excessive respiration, moisture content, infestation, light, and infestation-controlling methods. Insects that are selective to attack the embryo cause great losses in germination as compared to others. Seed loss can be detected by standard germination tests [31].

4.8. Commercial losses

Commercial losses are either due to direct consequences (foregoing factors) or indirect consequences (cost of preventive action or equipment). There may be a loss of goodwill, monetary loss, and loss caused by legal action. Intercountry trade might be affected by commercial loss. Losses can be rapidly reduced by experience and knowledge. Postharvest losses not necessarily take place due to inappropriate storage. There may be biological, physical, or mechanical factors in deterioration of cereal seeds. There is a need to broaden the intervention techniques to deliver high-quality product from the field to the market. For example, after the Tanzania outbreak of pest in the maize crop, Somalia and Malawi refused to take the maize due to insect spreading [32].

4.9. Temperature-dependent injury

High-temperature solar radiation on exposure deteriorates the fresh commodities rapidly. By increase of temperature, respiration also increases; there should be proper ventilation and cooling of the crops to obviate the problem. In similar way, there can be crop injury due to low temperature between 0 and −2°C. However some crops of tropical and subtropical origin showed tolerance for chilling injuries at 12–14°C. Chilling injuries (skin pitting, discoloration, uneven or abnormal ripening, and sensitivity to rapid deterioration) are apparent when commodity is removed from the environment [7, 29].

5. Assessment of losses

5.1. Assessment of seed losses in storage

The agents causing seed loss during storage are insects, rodents, and molds. In past this issue gained attention of many scientist but still there is a need to find some more effective technique to avoid seed loss due to insects. Insects can cause both qualitative and quantitative loss by boring or feeding on seeds; weight loss has got more attention [6].

5.2. Weight losses caused by insects

The assessment is made through collection of samples of seed and various intervals after storage and comparing the samples to observe the changes in subsequent samples. Assessing quantity loss with subsequent samples at different intervals will be used to estimate storage losses at different occasions. Sample collection from each batch of seed and their quantity loss are estimated accordingly. Sample should be collected without disturbing the pattern of infestation in the bulk stores. Three samples must be collected when subsequent regular sampling is not possible: first, at time of storage; next, in the midway of storage period; and last, at the final few weeks of seed storage.

The pattern of used seed and quantity loss is noted [22, 27].

5.3. Methods of estimation of weight loss

There are two methods used for estimation of weight loss by the insects, when subsequent sampling is possible: volumetric method and thousand grain mass (TGM) method.

In this case, when subsequent sampling is impossible, count and weight and converted percentage methods are used [6].

5.4. Volumetric method

Volumetric method is commonly known as the standard volume weight (SVW) or bulk density method. This is used to measure the bulk density of clean sample by the use of equipment. SVW is figured out from the sample of the seeds at the beginning of the storage period, and losses are determined. This method strictly calculates the weight loss by grain boring of insects and moisture difference over a subsequent time period in standard volume container. In stand volumetric method to determine accurate ratio for moisture content and dry weight of seed, moisture can be taken as constant term and the crop as dry matter. However, volume and frictional characters can also be affected by changes in moisture content. There is direct relation between moisture content and volume of the sample therefore seed should be packed loosely. To make the moisture constant, it is necessary to calculate the standard volume of dry matter at different moisture contents. The process is time-consuming and needs care and equipped laboratory [33]. Another factor affect the volume of sample is weight of insecticidal dust, adheres to the seed surface increase the volume of seed and frictional character. Sieving can be a useful phenomenon to remove dust. However, volumetric phenomena are less helpful due to overassessment of losses [6, 34].

5.5. Thousand grain mass method

This method differs from volumetric method in comparison with fixed number of seeds instead of fixed volume. This means that weight of seeds is multiplied by thousand and corrected by a dry matter. It is calculated by weighing and counting the seeds in a given sample. Standard reading is made by taking measurement at the beginning of storage of seeds, and then subsequent readings are compared with baseline reading [35].

5.6. Count and weight method

The method, sometimes called gravimetric method, is applied where the baseline readings of seed storage are not obtained in the start of the season. A sample of 1000 seeds and minimal medium are used in this estimation. Weight and number of seeds in each sample fraction is determined after isolation of damaged seeds. The results are then calculated by putting the values in the following equation:

$$\frac{(U \times N_d) - (D \times N_a)}{U(N_d + N_a)} \times 100 = \%\ \text{weight loss} \tag{1}$$

where U = weight of undamaged seeds, D = weight of damaged seeds, N_a = number of undamaged seeds, and N_d = number of damaged seeds.

In this method, moisture content of the separate fraction is unnecessary for a single sample, and differences of assumption are likely small. The method does consider hidden infestation in damaged category, and it considers seed infestation that is random by the insect which is not necessarily to be true [36]. The method can cause misleading results for low level of infestation and multiple infestations in large seeds. The method is useful in field level for quick estimation at extremes. To overcome the biased estimation, several refinements have been made. For example, different-sized seeds have different moisture contents and can have hidden infestation. These seeds are graded according to size and categories before counting and weighing [22]; severely attacked grains are separated, and reading of hidden grains is taken after emergence of infestation [37, 38]. There is another way to know the hidden infestation by dissecting seeds, but the method is tedious and has a chance of change in moisture content of seeds, for the sake of calculations that are needed to be made on dry matter.

5.7. A count and weight method modification

In this modification of count and weight method, highly infected grains have also been taken into account. To avoid the underestimation of losses, the number of infected seeds is counted for sum of the seeds in whole sample [39]. To assess the weight loss, the following equation is used:

$$\frac{\left[\mathrm{TND}(\mathrm{D}+\mathrm{U})\right]+\mathrm{FW}(\mathrm{N_d U}-\mathrm{N_a D})}{\left[\mathrm{TND}(\mathrm{D}+\mathrm{U})\mathrm{U}\right]+\left[\mathrm{FW}(\mathrm{N_d}+\mathrm{N_a})\mathrm{U}\right]}\times 100 \quad (2)$$

where U is the weight of undamaged seeds, D is the weight of damaged seeds, N_a is the number of undamaged seeds, and N_d is the number of damaged seeds.

The weight loss is calculated by averaging weight loss of two subsamples.

The percentage weight loss is calculated by deviation in Eq. (3):

$$\frac{\text{Undamaged weight (UW)} - \text{final weight (FW)}}{\text{Undamaged weight (UW)}}\times 100 \quad (3)$$

The undamaged weight of the sample (in the absence of missing seeds) is figured out by applying assumption in count and weight method. The average unit weight of undamaged grains in the original sample will be equal to the average unit weight in the remaining undamaged grain subsamples. The undamaged weight of the whole original grain sample is figured out as the product of the total number of seed estimated in the original sample and unit weight of undamaged seeds in the subsamples. The total number of seeds is the sum of the damaged and undamaged seeds. The final step will be the ratio of the final sample weight to the average unit weight of the seeds. The method focuses on missing seeds of the sample instead of underestimating lost samples as in conventional count and weight method [39].

5.8. The converted percentage damage method

The method is suitable for quick estimation of loss by boring insects without using equipment such as during a rapid field appraisal. Weight loss is assessed by taking percentage damage seed as sample. A study has been made to find out the relationship of damage and weight loss [23]. A conversion factor can be used subsequently to estimate the weight loss of same type to seeds in different samples. The conversion factor is calculated by the following formula by using same count and weight method and so faces same type of errors:

$$\frac{\%\ \text{damage seeds}}{\%\ \text{weight loss}} = \text{Conversion factor} \tag{4}$$

To overcome the error of count and weight method, 10% or more damaged sample seeds are used. The sample size should not be smaller than 500 seeds taken as the percentage of the total seeds present. Rapid loss technique is a rapid weight loss assessment in the field based on the damage and weight loss relationship. Standard graphs are used in this technique to relate percentage damage and weight loss of samples under study [6, 21]. For preparing reference graph, ten samples of 500 seeds are required in the laboratory. Weight loss is estimated by count and weight method, and percentage damage is calculated. The graph is plotted between percentage weight loss and percentage damage. In the field, clean sample of 200 seeds is gathered, percentage damage seed is determined, and loss figure of reference to graph is read off. For a sample of 200 seeds of Bambara or cowpeas, the technique depicts accuracy on ±7%. The method is better than percentage damage method and allows good assessment of weight loss by taking many individual samples in short time period at the field. Visual scales are rapid loss assessment methods. The abovementioned techniques are time-consuming and need appropriate equipment and well-trained technicians. Visual scale has benefit in assessing damage in fields, is ideal for field use and rapidity, and is required only for reference scales and operator unbiased. This technique enables wider coverage in small time and small sample error. On-site loss assessment avoids extra spoilage and anomalous results by double checking on the site before leaving. The visual scale should be calibrated according to the objective of the study, weight loss, and farmer perception of value [6, 21].

5.9. Losses by vertebra pests

Vertebra pests such as rodents and birds remove the whole seed from the sample so it is impossible to estimate loss by vertebra pests. The loss can be assessed by comparing the reference percentage of seed loss and average seed weight [40]. Population studies and feeding trials are used to calculate the losses by pests and rodents, but these are often with small accuracy in comparison to efforts expanded [41]. Pests used stored grains only as their part of diet; feeding trial can overestimate the loss of stored seeds. The quantity of seed loss by rodents is questionable. Loss of crops due to rodents goes behind when compared with loss of storage container, personal property, buildings, and potential health risks.

5.10. Weight loss by molds

Seeds infected by molds will cause weight loss; the weight loss can be assessed by the same method that is used to calculate weight loss by insect. The moldy seed weight loss increased due to the absorption of moisture so the loss by mold can be compensated. The method is not that good to assess the real loss of seeds, and the seeds may be misled as undamaged seeds due to no apparent indication of infestation on surface. To estimate the weight loss due to mold, damaged seeds are separated from undamaged seeds, and then moldy seeds are discriminated from damaged seeds. The weight loss due to mold will be equal to the weight of mold [21, 27].

5.11. Total loss in a season

The above losses are the seed losses at a given point in a time. The picture can be misleading; there must be a relation in the pattern of seed used in a season. In an undisturbed stored crop, if the sample loss is 10% throughout the storage seasons, the total loss will be due to insects. The seeds will loss at different quantities at different intervals of time due to insect exposure at different time lengths throughout the season [27]. The percentage loss of seeds increases gradually with time due to increase in insect infestation. After permitting changes in moisture content, the loss in seeds can be calculated by measuring the weight of seeds at the time when the seeds are still in the store and when taken out from the store. The loss other than insect damage can be obtained by subtracting the loss caused by other insects. The overall losses of the seed after store are considerably lower than the suggested value. Many loss assessment protocols at commercial and farm level have been reported [22]. The key to get the best assessment result is to draw an acceptable methodology for each commodity.

There were small figures for losses in commercial operations and none for cooperative level storage. The condition reflects the buying and selling of seeds in the developing world in short time period. This gives the picture of private sector (market liberation and parastatal marketing) involvement, but there is few information about storage loss. Entrepreneurs could store large quantity of seed for longer time period. However private sector has increased this level for farm storage. Many efforts, time and money was invested to measure the storage losses in farm storage, but the effort was not fruitful like the earlier projects. Moreover, the study should be undertaken with the postharvest sector as a whole, and precise measurements should be avoided. Social survey can be helpful to discover the farmers' problem for loss assessment and appropriate measurement techniques [42].

6. Harvesting and maturity indices

Harvesting and improper handling of commodities can cause bruises and injuries which directly affect their market value and make them unattractive. Injuries give the open space for microbial attack, that cause rotting, respiration increase, and shortage of storage life. Improper harvesting can cause crop loss and severe damage in seeds [9, 43].

6.1. Harvesting and handling

Harvesting is the first step of postharvest and is the last step of crop production. The method and condition of the harvest affect the further handling, processing, and storage of crops. Premature harvest causes loss of quality of seeds, and due to high water content, they will deteriorate in the store. Overmature crop harvest causes biological and physical losses of crops by consistent wetting and drying of crops [9]. Damp seeds need to be threshed quickly and dried after harvest. Different parts of plant are harvested by different harvesting methods; in case of forage, the whole plant is trimmed off; in case of cereal seeds, partial or part of the plant is threshed and cleaned; and straw or chaff is removed for further processing. Small scale produces performed threshing and harvesting by threshing combines harvesters (equipped through community groups) while in developing countries threshing and harvesting carried out by hand unlikely cause damage or deterioration of crops in store. Large-scale commercial producers use mechanical harvester equipment; their use is limited due to the production of cash crops. Post-harvest risk of crop damage in storage is reduces by manual harvesting Small scale produces performed threshing and harvesting by threshing combines [9, 21]. Conventionally, seeds are beaten with stick or against hard surface (wooden bar, log of wood, stone, and wooden metal or tub) for the sake of threshing. The methods can cause damage or cracks on the seeds, while seeds that are trodden under foot will be a less damaging method for the seeds. Grain heads or ears of sorghum, millet, or wheat are commonly beaten with sticks. However, manual harvesting is laborious, and economical process has the risk of physical damage. Maize cobs are beaten with sticks or shelled by hand, which result in high-level damage. Mechanized threshers are made to reduce the damage of seeds; the models are highly sophisticated.

Seed harvesting is carried out by a combination of various steps of threshing, cleaning, or combine harvester. Large-scale harvesting is carried out by mechanical equipment that is especially designed to harvest cereal seeds [6].

6.2. Storage structures for seeds

Seeds/grains are durable crops that usually require simple systems in their storage.

6.3. Farm-level storage

For protective outdoor or indoor storage, the seeds must be guarded for physical damage such as high temperature, adverse weather, snow and rain, and biological factors which include microorganisms, birds, rodent, mites, and insects. The process of farm storage is used by various countries to store major part of seeds [44, 45]. The storage structure varies in capacity and has range of 100 kg to few metric tons. Modifications in locally prepared storage structure could be done according to climatic conditions. There are some traditional storage structure. The bins are commonly made of ferro-cement, plywood, metal, and high-density and high-molecular-weight polyethylene. Plywood is the most suitable storage structure and underground structures of different shapes and sizes, which work on the principle of hermetic storage. This builds the higher level of carbon dioxide and low oxygen level which is lethal to pest and microbial attack during seed storage [46]. Traditional methods are cheaper, but they are not effective against pest and microbial attack. Silos or metal bins are also used at farm level to store seeds.

6.4. Bagged storage

In developing countries, seeds are stored in gunny or woven polypropylene bags in traditional warehouses, whereas silos for bulk storage, seeds elevator, and flat storage structure are used in developed countries [45, 47]. Bag storage is a laborious and costly process and has greater chance of seed spillage and biological losses. There may be water seepage and humidity problem due to inappropriate flooring of warehouses. Bags do not need any fumigation facilities or aeration system. In developing countries, the system will be uneconomical due to small farm size and cheaper manual labor.

6.5. Bulk storage

There are two types of bulk storage of seeds; it may be vertical (silos or bins) or horizontal (on floor stores). Horizontal stores comprise especially constructed floors of warehouses containing proper ventilation on the floor and walls that are strengthened to withstand the weight of the seeds. Bins and silos are especially designed stores, either round or square, in grouped or freestanding form, and incorporating unloading and loading usually of aeration systems. The alternative forms of bulk storage are belowground or partially belowground storage or enameled, sealed silos, for the storage of high moisture content seeds. The process is appropriate to bulk handling or storage of seeds [48, 49].

6.6. Hermetic storage

The traditional methods of storage in the natural build of oxygen and lower level of oxygen protect the seeds from biological damage. This traditional storage method is not effective for the seeds that have lesser moisture content, and infestation is less than insects per kilograms of seeds. The controlled atmosphere treatment and fumigation have to be supplemented in hermetic storage [50].

6.7. Outdoor storage

This is the temporary measure of storage in case of lack of permanent storage. The godowns and silos are built on plinth, and the stacks of seeds are covered with polyethylene covers. The stacks must be properly aerated in a week by raising the cover to the seventh or eighth layer. The cover and plinth (CAP) technique is used commonly for wheat and paddy. However, there is a chance of damage of cover by wind or rain, and effective fumigation cannot be achieved [44].

6.8. Storage practices for quality seeds

Quality of seeds can be maintained by especially designing the small stores to silo or warehouses which play a protective role against adverse temperature conditions, ground water, rain water, pests, and thefts. The structure and contents of the store should be managed [9].

7. Moisture control

A water disposal system and well-designed roof are required (overhung or gutter) to hinder the water running to the store. Drains move the water away from the stores. In large warehouses, side by side joining of stores should be avoided to prevent the water run into the store. Raised floor and proper drainage system protect against the risk of the flood, and water-proof floors and walls guard against ground water. There must be proper ventilation system to control the humidity inside the storage structure [27].

7.1. Protection against temperature

It is difficult to control temperature in the storage structures; it needs special design elements to control temperature. Temperature can be ascertained by the use of controlled ventilation. In the cold night, temperature can be manipulated by insulated stores. Stores can be built in east-west direction, and using shiny material outside the stores can be an effective strategy to control heat. Thick wall and wide roofs (for shade) can further reduce the heat of the storage structure. Heater and refrigerators can be installed to modify temperature in stores; the equipment will perform well in case of insulated stores. However in these stores, degree of insulation depends upon atmospheric condition outside of it. [6, 51].

8. Modified atmosphere storage (MAS) and controlled atmosphere storage (CAS)

In modified atmospheric condition, the gases are added or removed from surrounding to make a controlled atmospheric composition around seeds, which differs from air (78.08% nitrogen, 20.95% oxygen, and 0.03% carbon dioxide). This includes the reduction of oxygen and elevation of carbon dioxide concentration. CAS and MAS are different in level of control although CAS is more exact. The process is used to facilitate whole-store fumigation [52, 53].

8.1. Protection against theft

Security guard or fence may be needed to guard the store. Lights are installed to illuminate the stores [6].

8.2. Protection from birds and rodents

Bird and rodent contact can be hindered by avoiding the free spaces for the access to the seeds. Rodent guards are commonly tight-fitting shelves, or conventional methods could be tying of thorn bushes to the stores as barrier. Walls of the stores should be smooth, and roof should be rodent proof to avoid their contact inside the storage structure [54].

8.3. Protection against insect

The small size insects can easily enters the storage buildings therefore it is difficult to make insect-proof walls and roofs because they can enter from even small cracks. For protection against insects, smooth walls, controlled ventilation, cleaned stores, floor without cracks, and curved floor wall can be effective. For protection against termites, termite-resistant material should be used (concrete, stones, fire bricks, and termite-resistant wood) [55].

9. Transport

The transport of commodities from field to storage houses can produce some injury, which can later cause deterioration of produce. This main focus is to keep the produce dry and moisture-free. The risk of insect infestation of seeds from the contaminated container has insect residual. Mechanical injury is produced from vibration during transport, poor condition of vehicle and roads, poor driving, insecure container stacking, the use of unsuitable container, and careless handling. Overheating is produced from the sun or vehicle engine that results in moisture loss of produce, and this leads to natural breakdown and decay [6].

10. Quality and safety

The quality of a product defines the class, degree, excellence, or superiority of the crop. The quality of the durable commodity is the combination of characteristics, properties, and attributes that give the commodity value as the food or a source of next crop yield. Foreign material or high moisture content can affect the marketing quality of the crop. High moisture may encourage the shrinkage or biological and biochemical damage in seeds. Low moisture in pulses and paddy rice can break or damage the seeds. The discolored and broken seeds lower the marketing quality and increase the chance to insects and microorganism attack [14, 27]. The main objective of a farmer is to produce apparently good product with few visual defects, and product must score high on yield, disease resistance, ease of harvest, and shipping quality and achieve the national and international quality standards. For the buyer or consumer, the appearance is more important; they are eagerly concerned in healthy seed and lifelong storage. The product should be assured safe for the distribution to suppliers and market [7].

11. Postharvest quality assessment techniques

There are two methods to assess the quality of crop after harvest such as analytical or objective and subjective or sensory method [56]. There are limitations for subjective method in quality evaluation of hazardous material [57]. Objective method is destructive or nondestructive in nature. Manual sorting of commodities is outdated, costly, unreliable, subjective, laborious, and slow. Food depicts complex and dynamic behavior with respect to varietal and time dependency. Nondestructive methods classified according to physical principles are visual inspection, x-ray imaging and computed tomography, near-infrared spectroscopy, ultrasonic testing, thermographic testing, electromagnetic testing, liquid penetrant testing, magnetic particle testing, acoustic emission testing, infrared and thermal testing, magnetic resonance imaging, electronic nose, etc. [58].

Quality assessment or grading has the following benefits:

i. It helps to facilitate the purchase of an unseen product.

ii. Quality and safety improvement incentives.

iii. Grading makes the market information meaningful.

iv. It helps in facilitating quality and price comparison.

v. It reduces the deception and fraudulent marketing risk.

vi. Diverse marketing mechanism can be enabled through the process such as commodity exchange, future trading, inventory credit, credit letters, and facilitating resolution of disputes regarding composition or quality [6].

12. Future challenges for postharvest technology

In increasing urban population and shifting lifestyle in developing countries, demand for food has increased many times. Seed is a unit which not only is used as food but also responsible for production of the next generation. The most attention-demanding perspectives in seed biology are postharvest management and quality maintenance of seed. Although in past years, great success has been achieved in development of novel packaging, storage and transport systems, pests, and seed-borne disease control for market access. However, to achieve future goals, further research and technology development should be dedicated to explore genetic aspects of quality traits like stress resistance, resistance to postharvest diseases, and pest management. Researchers should try to work on integrated approaches for postharvest management of seed. Nanotechnology is an emerging field which is leading to tremendous achievements in crop sciences. Seed biologist should also try to explore this field so that seed could be preserving for longer time without altering the genetics.

13. Conclusions

Seed quality should be maintained for the better quality crops. Maintenance of the seeds is the major problem in the developing countries these days. There is a need to establish better, economical, convenient, and productive methods for postharvest handling and seed storage. Researchers should focus on translating knowledge into agricultural outputs. The chapter deals with the value of healthy seeds, factors affecting the seed quality, postharvest techniques for the seed storage, and techniques and safety measures for their quality assessment for the maintenance of good quality seeds to meet the international standards emphasizing on the needs of developing countries. This could be achieved by opting more sensitive and advance technologies with special emphasis on seed response to environmental and maternal factors.

Author details

Zubaida Yousaf*, Nadia Saleh, Asba Ramazan and Arusa Aftab

*Address all correspondence to: z4zubaida@yahoo.com

Department of Botany, Lahore College for Women University, Lahore, Pakistan

References

[1] Delouche JC, Caldwell WP. Seed vigor and vigor tests. Proceeding of Association of Official Seed Analyst. 1960;50:124–129.

[2] Delouche JC. Planting seed quality. In: Proceeding of 1969 Beltwide Cotton Prod.-Mech. Conference. New Orleans. January 1969. p. 16–18.

[3] Grabe DF. Agronomic significance of seed deterioration. Agronomy Abstract. 1965;57:40.

[4] Woodstock LW. A respiration test for corn seed vigor. Proceeding of Association of Official Seed Analyst. 1966;56:95–98.

[5] Gregg BR, Billups GL. Seed Conditioning Technology Part A. Science Publishers. USA. 2010, Vol. 2, p. 1–2.

[6] Boxall RA, Brice JR, Taylor SJ, Bancroft RD. Crop post-harvest, science and technology, volume 1, principles and practice. In: Golob P, Farrell G, Orchard E (Eds) Crop Post-Harvest: Science and Technology. Blackwell Science Ltd, Oxford, UK. 2002. p. 141–204. doi: 10.1002/9780470751015.ch5.

[7] Mohammed M. Manual on post-harvest management strategies to reduce losses of perishable crops. UWI/CTA/NAMDEVCO Workshop on Strategies to Reduce Post-

harvest Losses in Perishable Commodities at NAMDEVCO Conference Facility. Piarco, Trinidad. February 24–25, 2014.

[8] Harrington JF. Seed storage and longevity. Cited in: Kozlowski TT, Seed Biology. 1972;3:145–245.

[9] Kiaya V. Post-harvest Losses and Strategies to Reduce Them. New York: Action Contre la Faim (ACF International). 2014.

[10] López J, Uribe E, Vega-Gálvez A, Miranda M, Vergara J, Gonzalez E, Di Scala K. Effect of air temperature on drying kinetics, vitamin C, antioxidant activity, total phenolic content, non-enzymatic browning and firmness of blueberries variety O Neil. Food and Bioprocess Technology. 2010;3:772–777.

[11] Mujumdar AS, Law CL. Drying technology: Trends and applications in post-harvest processing. Food and Bioprocess Technology. 2010;3:843–852.

[12] Chen XD, Mujumdaer AS (Eds) Drying Technologies in Food Processing. Oxford: Blackwell. 2008.

[13] Boothumjinda S, Exell RHB, Rongtawng S, Kaewnikom W. Field tests of solar rice dryers in Thailand. In: Proc ISES Solar World Forum. Perth, Oxford: Australia (Parkville ISES), Pergamon Press. 1983. p. 1258–1263.

[14] Kitinoja L, Gorny JR. Post-harvest Technology for Small-scale Produce Marketers Economic Opportunities, Quality and Food Safety. Post-harvest Horticulture Series. Department of Pomology, University of California. Davis, USA. 1999. p. 21.

[15] Radajewski W, Jolly P, Abawi GY. Optimization of solar grain drying in a continuous flow dryer. Journal of Agricultural Engineering Research, 1987;38:127–144.

[16] Colliver DG, Brook RC, Peart RM. Optimal management procedure for solar grain drying. American Society of Agricultural Engineers. Chicago. 1978;12:78–35.

[17] Heid WG. Solar-assisted Combination Drying: An Economic Evaluation. Agricultural Economic Report, Economics, Statistics and Co-operative Services, US Department of Agricultural, No. AER-453. 1980.

[18] Hellickson MA. Cost/performance goals for agricultural solar collectors. American Society of Agricultural Engineers. 1979;54:79–40.

[19] Chung DS. Lee C. Physical and thermal properties of grains. In: Champ BR and Highley E (Eds) Preserving Grain Quality by Aeration and In-store Drying. Proceedings of an International Seminar. Kuala Lumpur, Malaysia. Canberra, Australia: ACIAR. October 9–11, 1986.

[20] Bartosik RE. A model-based fan and burner control strategy for the in-bin drying and conditioning of corn. 2005.

[21] Harris KL, Lindblad CJ. Post-harvest Grain Loss Assessment Methods. Minnesota: America Association of Cereal Chemist. 1978. p. 193.

[22] Boxall RA. A Critical Review of the Methodology for Assessing Farm-level Grain Losses after Harvest. Slough, UK: Tropical Products Institute G191. 1986.

[23] Boxall RA. Storage losses. In: Golob P, Farrell G, Orchard JE (Eds) Crop Post-harvest: Science and Technology Volume 1: Principles and Practice. Oxford: Blackwell Sciences Ltd. 2002. p. 143–169.

[24] Kader AA. Post-harvest Technology of Horticultural Crops. 3rd ed. Oakland: University California Agriculture and Natural Resources, Publ. 3311. 2002.

[25] Gross K, Wang CY, Saltveit ME. The Commercial Storage of Fruit, Vegetables and Florist and Nursery Stocks. USDA Agriculture Handbook 66. 2002. http://www.ba.ars.usda.gov/hb66/index.html.

[26] Atanda SA, Pessu PO, Agoda S, Isong IU, Ikotun I. The concepts and problems of post-harvest food losses in perishable crops. African Journal of Food Science. 2011;5:603–6013.

[27] Boxall RA. Post-harvest losses to insects – A world overview. International Bioterioration and Biodegradation. 2001,48:137–152. Retrieved from www.scopus.com.

[28] Grolleaud M. Post-harvest Losses: Discovering the Full Story. Rome: FAO. 1997. Available from: http://www.fao.org/docrep/004/ac301e/AC301e04.htm#3.2.1%20Rice

[29] Lipinski B, Hanson C, Lomax J, Kitinoja L, Waite R, Searchinger T. Reducing Food Loss and Waste. World Resources Institute Working Paper. Washington, DC: 2013.

[30] Grolleaud M. Post-harvest Losses: Discovering the Full Story. Overview of the Phenomenon of Losses during the Post-harvest System. Rome, Italy: FAO, Agro Industries and Post-Harvest Management Service (AGSI). 2002.

[31] ISTA. International rules for seed testing. Proceedings of the International Seed Testing Association. 1966;31:49–91.

[32] Tyler PS, Golob P, Compton J, Bickersteth S. Study on Phytosanitary Requirements to Promote Maize Trade in Eastern and Southern African Countries at Risk from the Larger Grain Borer. Report for the European Commission. Chatham, UK: Natural Resources Institute. 1990.

[33] Adams JM, Schulten GGM, Harris KL, Lindblad CJ. Post-harvest Grain Loss Assessment Methods. American Association of Cereal Chemists. USA. 1978 p. 83–95.

[34] FAO I. WFP. The State of Food Insecurity in the World. Food and agriculture organization of the united nations. Rome. 2013.

[35] Reed C. The precision and accuracy of the standard volume weight method of estimating dry weight losses in wheat, grain sorghum and maize, and a comparison with

the thousand grain mass method in wheat containing fine material. Journal of Stored Products Research. 1987;23:223–231.

[36] Adams JM, Harman GW. The Evaluation of Losses in Maize Stored in a Selection of Small Farms in Zambia with Particular Reference to the Development of Methodology. Slough, UK: Tropical Products Institute G109. 1977.

[37] Ratnadass A, Fleurat-Lessard F. Improvement in the practical methods of assessment of losses caused by insects in grain stored at the village level in tropical Africa. In: Fleurat-Lessard F, Ducom P (Eds) Proceedings of the 5th International Working Conference on Stored Product Protection, Bordeaux. Bordeaux, France: INRA. 1991.

[38] Ratnadass A, Berté S, Diarra D, Cisé B. Insect losses in sorghum stored in selected Malian villages, with particular emphasis on varietal differences in grain resistance. In: Highley E, Wright EJ, Banks HJ and Champ BR (Eds) Proceedings of the 6th International Working Conference on Stored Product Protection, Canberra. Wallingford, UK: CAB International. 1994.

[39] Compton JAF, Floyd S, Ofosu A, Agbo B. The modified count and weigh method: An improved procedure for assessing weight loss in stored maize cobs. Journal of Stored Products Research. 1998;34:277–285.

[40] Boxall RA, Gillett R. Farmer Level Storage Losses in Eastern Nepal. Slough, UK: Tropical Products Institute G157. 1982.

[41] Hernandez A. Drummond DC. A study of rodent damage to food in some Cuban warehouses and the cost of preventing it. Journal of Stored Products Research. 1984;20:83–86.

[42] Goletti F, Wolff C. The Impact of Post-harvest Research. Washington, DC: International Food Policy Research Institute, Markets and Structural Studies Division Discussion Paper, 29. 1999.

[43] FAO. Global Food Losses and Waste: Extent, Causes and Prevention. . Food and agriculture organization of the united nations Rome. 2011.

[44] Semple RL. Post-harvest technology: Storage development and application in developing Asian countries. In: Semple RL, Hicks PA, Lozare JV, Castermans A (Eds) Grain Storage Systems in Selected Asian Countries. REAPASIA. People Republic of China. 1992. p. 9–59.

[45] McFarlane JA. Storage methods in relation to post-harvest losses in cereals. Insect Science Application. 1988;9:747–754.

[46] Shejbal J, de Bioslambert JN. Modified atmosphere storage of grains. In: Multon JL (Ed). Preservation and Storage of Grains, Seeds and Their By-Products: Cereals, Oilseeds, Pulses, and Animal Feed. New York: Lavoisier. 1988. p. 749–777.

[47] Kennedy L, Devereau AD. Observations of large-scale outdoor maize storage in jute and woven polypropylene sacks in Zimbabwe. In: Highley E, Wright EJ, Banks HJ,

Champ BR (Eds) Stored Product Protection: Proceedings of the 6th International Working Conference on Stored-Product Protection. Wallingford: CAB International. 1994. p. 290–295.

[48] Multon JL. Spoilage mechanisms of grains and seeds in the post-harvest ecosystem, the resulting losses and strategies for the defence of stocks. In: Multon JL (Ed.) Preservation and Storage of Grains, Seeds and Their By-Products: Cereals, Oilseeds, Pulses and Animal Feed. New York: Lavoisier. 1988. p. 3–63.

[49] Bailey JE. Whole grain storage. In: Sauer DB (Ed). Storage of Cereal Grains and Their Products. St. Paul, MN: American Association of Cereal Chemistry. 1992. p. 157–182.

[50] Alvindia DG, Caliboso FM, Sabio GC, Regpala AR. Modified atmosphere storage of bagged maize outdoors using flexible liners: A preliminary report. In: Highley E, Wright EJ, Banks HJ, Champ BR (Eds) Stored Product Protection: Proceedings of the 6th International Working Conference on Stored-Product Protection. Wallingford: CAB International. 1994. p. 22–26.

[51] Longstaff BC. Temperature manipulation and the management of insecticide resistance in stored grain pests: A simulation study for the rice weevil, Sitophilus oryzae. Ecological Modelling. 1988;42:303–313.

[52] Banks HJ, Annis PC. Comparative advantages of high CO_2 and low O_2 types of controlled atmospheres for grain storage. In: Calderon M, Barkai-Golan R (Ed.) Food Preservation by Modified Atmospheres. Boca Raton: CRC Press. 1990. p. 94–122.

[53] Paster N, Calderon M, Menesherov M, Barak V, Mora M. Application of biogenerated modified atmospheres for insect control in small grain bins. Tropical Science. 1991;32:355–358.

[54] Smith RH. Rodents and birds as invaders of stored-grain ecosystems. In: Jayas DS, White NDG, Muir WE (Eds) Stored-Grain Ecosystems. New York: Marcel Dekker. 1995. p. 289–323.

[55] Hagstrum DW, Flinn PW, Shuman D. Acoustical monitoring of stored-grain insects: An automated system. In: Highley E, Wright EJ, Banks HJ, Champ BR (Ed.) Stored Product Protection: Proceedings of the 6th International Working Conference on Stored-Product Protection. Wallingford: CAB International. 1994. p. 403–405.

[56] Ranganna S. Handbook of Canning and Aseptic Packaging. Tata McGraw-Hill Publishing Company. Dehli, India. 2000.

[57] Jha SN, Garg R. Non-destructive prediction of quality of intact apple using near infrared spectroscopy. Journal of Food Science and Technology. 2010;47: 207–213.

[58] Siddiqui MW. Post-harvest Biology and Technology of Horticultural Crops: Principles and Practices for Quality Maintenance. CRC Press. USA. 2015.

Chapter 6

Designing Novel Breeding Strategies for Producing High-Oil Crops Based on a Molecular Understanding of Triacylglycerol Metabolism

Masatake Kanai, Shoji Mano, Makoto Hayashi and Mikio Nishimura

Additional information is available at the end of the chapter

http://dx.doi.org/10.5772/64465

Abstract

Seeds are storage organ in plants and main resource of plant oils to human civilization and the demand of plant oils are increasing yearly and expansion of the production capacity is an urgent issue worldwide. Thus, it is necessary to improve oil yields per unit area and generation of crops with high-oil content is needed. *Arabidopsis thaliana* plays a vital role in advancement of genetics and molecular biology in plant sciences. The forward and reverse genetic approaches with Arabidopsis have provided an overview of triacylglycerol metabolism. The elucidation of the overview contributes to understanding of spatiotemporal regulation of a metabolic flow of triacylglycerol metabolism in plant cell. This understanding sheds light on bottlenecks in triacylglycerol biosynthesis and provides novel clues for increasing seed triacylglycerol content. Recent advance in metabolic engineering approaches demonstrate several evidences that triacylglycerol metabolism is coordinated with other metabolisms. Most notably, triacylglycerol biosynthesis competes with biosynthesis of starch or seed storage proteins. These studies indicate that alterations of the metabolic pathways to avoid the competitions could be a novel concept for increasing seed oil content.

Keywords: Seed, oil, Triacylglycerol, Metabolic engineering, *Arabidopsis thaliana*

1. Introduction

Seeds are storage organs in plants that accumulate massive quantities of carbohydrates, proteins, and oils, which are collectively referred to as seed storage reserves. Seed storage reserves are utilized to supply energy and nutrients to the embryo during postgerminative growth until the plant acquires the ability to perform photosynthesis. Hence, seed storage reserves are an easily degradable source of organic matter for organisms.

Seeds are widely used as industrial materials. For example, plant oils are mainly derived from seeds. In 2014, the overall consumption of plant oils was 170,980 kt, which is nearly double that consumed in 2004. This increase is attributed to a greater demand for dietary oil, as well as industrial materials for use as carbon-neutral oils, representing an alternative to petroleum.

To improve oil productivity in plants, it is essential to increase seed yields in crops. However, the agricultural area worldwide has been flat for 40 years. Therefore, improving seed yield per plant has become increasingly important. Since increasing seed yield is one of the major issues in plant science, effective strategies for increasing yield have been investigated. Considering the many successes in the field of metabolic engineering of microorganisms, metabolic engineering represents a promising approach for increasing oil yields in seeds. Lipid metabolism has been extensively studied, and its metabolic pathways and regulatory systems have been elucidated. Additionally, in-depth analysis of crop genomes has been greatly expedited by recent advances in life science technologies (next-generation sequencing technology, genome editing, and so on). Therefore, numerous translational studies of model plant species have been performed, which have shed light on crop species. This chapter summarizes recent advances in our understanding of oil metabolism in seeds and introduces promising strategies for increasing oil production in crops.

2. Main components of seeds

Seeds accumulate carbohydrates, proteins, and oils as seed storage reserves. The ratios of these components dramatically differ among plant species, representing a major factor that determines the usage and application of seeds [1].

2.1. Cereal seeds

Starch consists of carbohydrates, which comprise polysaccharides composed of glucose [2–4]. Major cereal seeds, such as rice, wheat, and maize, contain large amounts of starch. Barley and rye seeds also accumulate starch as their main storage reserves. The starch contents in these cereal seeds are over 70%, whereas the contents of proteins and oils are less than 20% (**Table 1**). Therefore, cereal seeds represent an important energy source for animals.

2.2. Oil seeds

Oil seeds mainly accumulate lipids. Oil seeds are one of the major sources of plant oils. These seeds are compressed and heated to extract oils from their cells. Oil crops, which produce oil seeds, are cultivated worldwide. Oil palm, soybean, and rapeseed, the "big three", represent three-quarters of total plant oil production (**Table 2**). Palm oil and olive oil are extracted from fruits, whereas many other oils are extracted from seeds (**Table 2**).

Grain	Carbohydrates	Proteins	Oils
wheat	72.2	10.6	3.1
rice	73.8	6.8	2.7
maize	70.6	8.6	5.0
barley	72.1	10.9	2.1
rye	70.7	12.7	2.7

Table 1. Carbohydrate, protein, and fat contents in cereal seeds. Values (%/w) in this table were extracted from the Food Composition Database of the Ministry of Education, Culture, Sports, Science, and Technology in Japan (http://fooddb.mext.go.jp/index.pl).

	Plant species	Production (kilotons, 2013–2014)	Rate (% total plant oil production)	Storage organ
1	oil palm	59,360	34.09	fruit
2	soybean	44,439	25.52	seed
3	rapeseed	27,029	15.52	seed
4	sunflower	16,169	9.29	seed
5	cotton	4,859	2.79	seed
6	peanut	3,548	2.04	seed
7	olive	3,416	1.96	fruit
8	corn	3,148	1.80	seed

Table 2. Major oil crops and worldwide oil production in 2013–2014. The rates of total production and % total plant oil production are summarized for major oil crops; the values were obtained from FAOSTAT (http://faostat.fao.org/).

Species	Concentration (%)		Variety
	Oil	Protein	
soybean	20.1	41.4	Kariyutaka
rapeseed	41.0	25.6	Kizakinonatane
sunflower	33.8	30.4	Mammoth Russian
peanut	42.6	25.8	Omasari
sesame	49.9	21.6	–

Table 3. Oil and protein concentrations in major oil seed crops. The oil and protein concentrations were measured following the methods of Kanai et al. (2016). The variety of sesame used in this experiment was not identified.

2.2.1. *Oil contents of oil seeds*

Of the major oil crops, the oil content of oil palm fruit is over 50%, whereas that of soybean and rapeseed is 20.1% and 41%, respectively (**Table 3**) [5, 6]. The seed oil content in sunflower, peanut, and sesame is 33.8%, 42.6%, and 49.9%, respectively (**Table 3**). Thus, the seed oil contents of typical oil crops range from approximately 20–50%. These values are relatively low compared to the starch contents of rice, wheat, maize, and so on, suggesting that there is strong potential for increasing oil contents in seeds.

2.2.2. Protein contents of oil seeds

The protein content of major cereal seeds is less than 15% (**Table 1**), whereas the protein content of soybean and rapeseed, which are major suppliers of plant oils, is 41.4% and 25.6%, respectively (**Table 3**). Most oil seeds accumulate more proteins, i.e., seed storage proteins (hereafter referred to as SSPs), than cereal seeds.

3. Triacylglycerol metabolism in seeds

Lipids are essential components not only for animals, but also for plants. Various species of lipids are biosynthesized throughout plant organs. Some lipids, such as membrane lipids [7], cuticular waxes [8], and volatile oils, comprising fatty acids [9–11], alcohols, terpenes, and so on, are produced in plants in response to changes in the external environment. On the other hand, seed oils, the primary subject of this chapter, are composed of triacylglycerol (referred to hereafter as TAG) [12–14]. Except for palm oil, major seed oils are liquids at room temperature (**Figure 1A**). One molecule of TAG is composed of a glycerol backbone and three fatty acids (**Figure 1B**).

3.1. Triacylglycerol biosynthesis

TAG biosynthesis primarily involves two steps, i.e., fatty acid biosynthesis and the assembly of fatty acids with a glycerol backbone (**Figure 1C**).

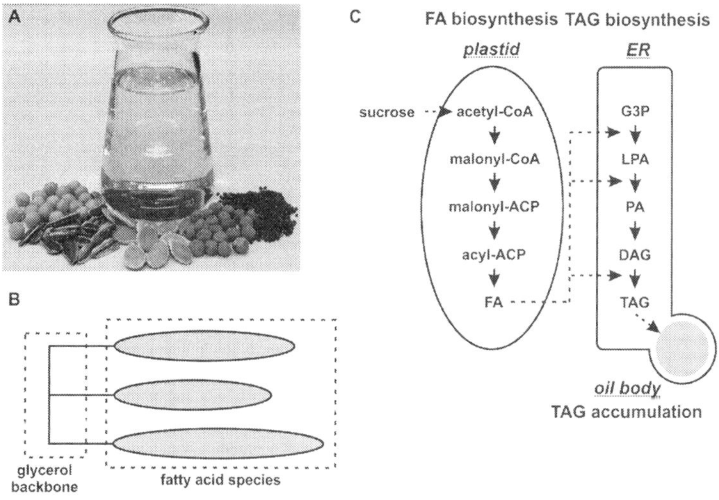

Figure 1. Triacylglycerol in plant seeds. (**A**) Plant oil extracted from seeds. (**B**) Structure of TAG. TAG is produced by the assembly of one glycerol with three fatty acids. (**C**) Schematic diagram of TAG biosynthesis in plant cells.

3.2. Fatty acid biosynthesis

In plant cells, fatty acids are synthesized in plastids. Fatty acid biosynthesis begins with the carboxylation of acetyl-CoA to malonyl-CoA by acetyl-CoA carboxylase (reaction (1) in **Figure 2**). The malonyl group in malonyl-CoA is transferred to acyl carrier protein (hereafter referred to as ACP), producing malonyl-ACP (reaction (2) in **Figure 2**). Next, 3-keto-butyryl-ACP is synthesized via the condensation reaction of malonyl-ACP with acetyl-CoA (reaction (3) in **Figure 2**). The 3-keto-butyryl-ACP molecule is converted to butyryl-ACP via reduction and dehydration (reaction (4) in **Figure 2**). The provision of C2 units, reduction, and dehydration are repeated, leading to the elongation of the carbon chain of acyl-ACP (carbon chain elongation in **Figure 2**). Synthesized acyl-ACP is catalyzed by thioesterase to form free fatty acids, which are again converted to acyl-CoA and imported to the endoplasmic reticulum (hereafter referred to as ER).

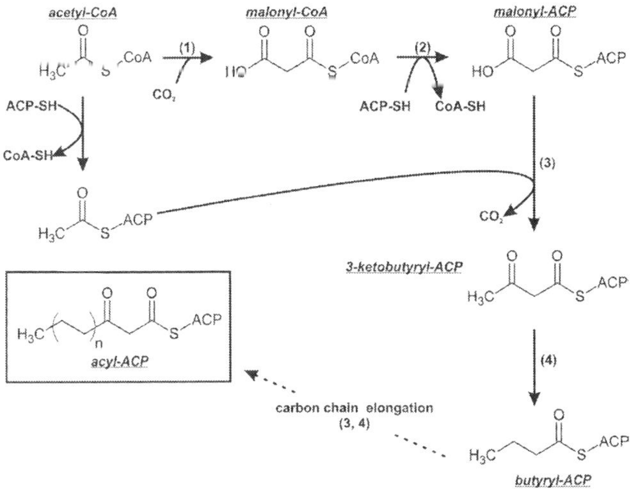

Figure 2. Fatty acid biosynthesis in plastids: (1) acetyl-CoA carboxylase, (2) malonyl CoA:ACP transacylase, (3) 3-ketoacyl-ACP synthase, and (4) 3-ketoacyl-ACP reductase, 3-hydroxyacyl ACP dehydrase, and enoyl-ACP reductase.

3.3. Assembly of fatty acids with the glycerol backbone

Acyl-CoA is a donor of the acyl-group for TAG biosynthesis. *De novo* synthesis of TAG begins with the transfer of the acyl-group to the sn-1 position of glycerol 3-phosphate, leading to the production of lysophosphatidic acid (reaction (1) in **Figure 3**). Subsequently, phosphatidic acid is produced by the transfer of the acyl group to the sn-2 position (reaction (2) in **Figure 3**). Next, the sn-3 position of phosphatidic acid is dephosphorylated and converted to diacylglycerol (reaction (3) in **Figure 3**), and, finally, TAG is synthesized by the transfer of the acyl-group to the sn-3 position of diacylglycerol (hereafter referred to as DAG) (reaction (4) in **Figure 3**). In addition to the *de novo* synthesis pathway, TAG is produced via an alternative pathway through phosphatidylcholine (hereafter referred to as PC) [15, 16]. PC, one of the main components of membrane lipids, is present in pools in the ER [16]. PC pools affect *de novo* TAG synthesis and the acyl-CoA pool in the cytosol to supply DAG as a precursor for TAG [16].

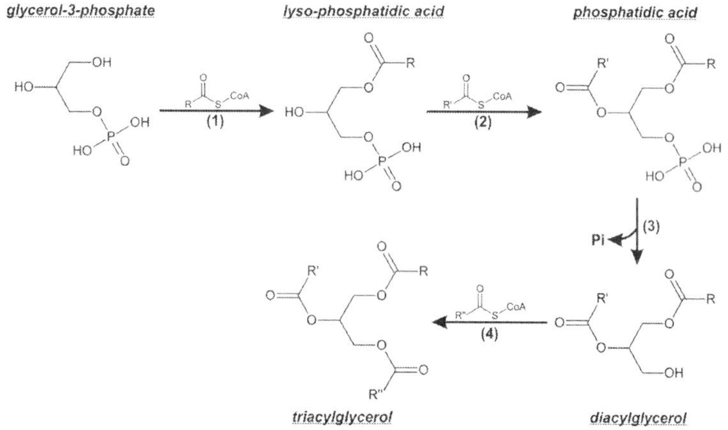

Figure 3. Triacylglycerol formation in the ER: (1) acyl-CoA:G3P acyltransferase, (2) acyl-CoA:LPA acyltransferase, (3) PA phosphatase, and (4) acyl-CoA:DAG acyltransferase.

3.4. Triacylglycerol accumulation

Synthesized TAGs accumulate in compartments in the ER (**Figure 1C**), which are converted to vesicles through budding. The vesicles then develop into oil-accumulating organelles, i.e., oil bodies [16–19]. Oil bodies are TAG storage organelles with a single layer membrane, whose major membrane protein is oleosin [18, 20, 21]. Oleosin blocks adhesion between neighboring oil bodies, which allows small oil bodies to be packed tightly together without adhesion in the cells of oil seeds [22].

4. Strategies for improving oil content by modifying triacylglycerol metabolism in oil seed crops

Since plant oils are commercially important, improving oil seed crops has long been a focus of breeders. Such breeding efforts, which began in ancient times, have led to the improvement of oil contents in several crops. In addition, the oil contents of the seeds of modern cultivars are significantly higher than those of wild species [23, 24]. Currently, many plant breeders have undertaken the challenge of further increasing seed oil contents. However, recent studies using quantitative trait loci analyses revealed that seed oil contents are controlled by many genes with additive effects [25–28], suggesting that traditional breeding methods based on cross-fertilization may be inadequate for further increasing seed oil contents.

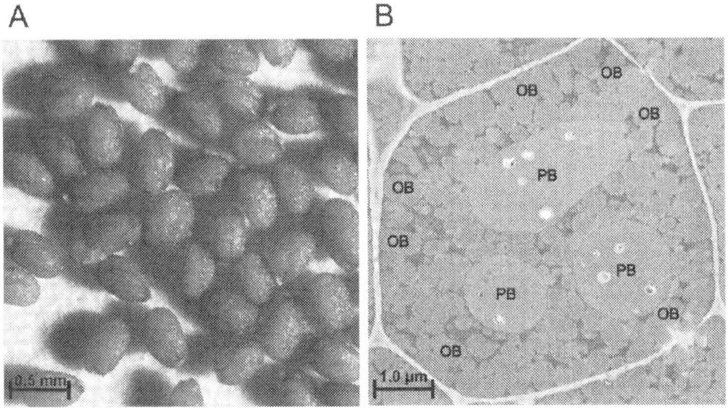

Figure 4. Seeds of *Arabidopsis thaliana*. (A) Microscopic image of Arabidopsis seeds. (B) Electron microscopic image of Arabidopsis seed cells. OB: oil body, an oil storage organelle in seeds. PB: protein body, an organelle that accumulates seed storage protein.

By contrast, genetic transformation may be an effective tool for increasing seed oil contents. The model plant *Arabidopsis thaliana*, a close relative of the major oil crop *Brassica napus*, produces typical oil seeds (**Figure 4A** and **B**). The use of Arabidopsis drastically simplifies both the processes of screening mutants with abnormal phenotypes and generating transformants, in which genes of interest are introduced. The use of Arabidopsis has accelerated the process of uncovering TAG metabolic pathways in plants. In fact, most genes encoding enzymes involved in TAG metabolism have been identified based on characterization of Arabidopsis mutants defective in TAG metabolism [29, 30]. Elucidating TAG metabolic pathways has revealed the limiting factors and key enzymes in this process and has led to the development of novel strategies for improving seed TAG contents. In this section, we review important strategies for improving TAG contents in seeds based on metabolic engineering approaches.

4.1. Enhancement of TAG biosynthesis

The TAG biosynthetic pathway, which extends across plastids, the cytosol, and the ER, involves many enzymes [30]. Additionally, DAG biosynthetic pathways include the *de novo* synthetic pathway and the phosphatidylcholine-derived pathway (hereafter referred to as PC-derived pathway), whose activities greatly differ among plant species [15, 16]. Thus, increasing the activity of enzymes in the *de novo* or PC-derived pathway does not always increase seed oil contents in every plant species. On the other hand, TAG production from DAG by acyl-CoA:DAG acyltransferase is a common pathway among plants. Overexpressing acyl-CoA:DAG acyltransferase significantly increases seed TAG contents in Arabidopsis [31] and in other plants species [32–37]. These findings indicate that the conversion of DAG to TAG is one of the rate-limiting steps in the production of TAG in seeds, and they suggest that the acyl-CoA:DAG acyltransferase gene would be a promising target gene for increasing TAG contents.

4.2. Suppression of TAG degradation

Synthesized TAG is stored in oil bodies and degraded during germinative growth [17, 38, 39]. The TAG degradation pathway has also been uncovered, and most genes encoding enzymes in this pathway have been identified [17, 39]. The expression of these genes is upregulated after seed imbibition, and TAG degradation activity rapidly increases in imbibed seeds [40–42]. These genes are also expressed during seed development in several plants [43, 44]. In fact, TAG degradation occurs in developing seeds [45–47]. Therefore, the TAG degradation pathway is activated during seed development, and seeds lose some of the TAG synthesized during seed development. This finding suggests that suppressing TAG degradation would be a promising strategy for improving seed oil contents. Oil degradation begins with TAG hydrolysis via TAG lipase. TAG lipase was genetically identified as *SUGAR DEPENDENT 1* in Arabidopsis [48] and was subsequently identified in rapeseed and Jatropha [47, 49]. Suppressing *SUGAR DEPENDENT 1* expression significantly increases seed oil contents [47, 49]. These reports indicate that suppressing TAG degradation via suppressing *SUGAR DEPENDENT 1* expression may represent an effective strategy for increasing seed oil contents in oil seed crops.

5. Novel strategies focused on carbon flux in seeds

In production systems that utilize microorganisms and cultured cells for metabolic engineering, individual cells synthesize organic compounds, providing a limitless source of carbon in the medium. By contrast, the seed, representing only one of many plant organs, synthesizes storage reserves using a limited carbon source provided by photosynthesis. This limited carbon source is thought to represent another limiting factor to the use of seeds as production systems, in addition to the limited activities of metabolic enzymes. To utilize seeds for oil production, it is important to obtain a comprehensive understanding of the factors that regulate TAG production, including carbon flow, metabolite transport, TAG metabolism, compartmentation, competition between other reserves, and so on. Recent studies identifying other factors that limit oil production in seeds have opened up the possibility of developing novel strategies for improving seed oil contents.

5.1. Effective utilization of "the window" during seed development

Fertilized embryos grow into mature seeds through sequential development. Seed storage oils are produced only during a short period of seed development; high TAG biosynthesis activity only occurs for a short period of time. The master transcription factor of TAG biosynthesis, WRINKLED1 (hereafter referred to as WRI1), is expressed during this short period and induces the expression of genes related to fatty acid biosynthesis [50–54]. Although WRI1 positively regulates TAG biosynthesis, the overexpression of *WRI1* fails to increase TAG contents in seeds

[51, 53]. These findings suggest that, in order to increase seed TAG contents, it is necessary to reconsider the timing and duration of *WRI1* expression. Arabidopsis seeds initially accumulate TAG, followed by proteins [1]. A detailed analysis of seed development revealed that there is a time lag between the termination of TAG biosynthesis and the initiation of protein biosynthesis, i.e., there is a "window" period in the middle phase of seed development during which the activities of the TAG and protein synthesis pathways are low (**Figure 5**) [54]. Additionally, strong expression of *WRI1* during this window extends the duration of TAG biosynthesis and increases TAG contents in seeds [54], indicating that *WRI1* expression during this window effectively induces the TAG biosynthesis pathway. This finding has important implications for developing novel strategies for increasing seed TAG contents. The timing and duration of TAG and seed storage protein synthesis differ greatly among plant species. Therefore, identifying the window phase and fine tuning of *WRI1* expression are essential for generating oil crops with increased TAG contents in seeds.

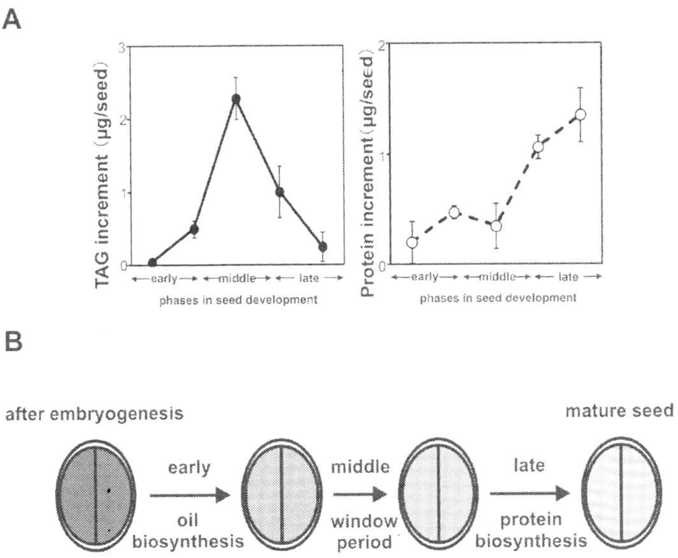

Figure 5. Oil and protein biosynthesis during seed development. (A) Phases of oil and protein biosynthesis during seed development in *Arabidopsis thaliana*. (B) Schematic diagram of oil and protein biosynthesis during seed development in *Arabidopsis thaliana*.

5.2. Concentrating the carbon source into triacylglycerol biosynthesis by reducing the levels of other organic materials

Seeds accumulate various organic materials, including carbohydrates and proteins, in addition to TAG. Therefore, reducing carbohydrate and protein levels would help direct the carbon source into TAG biosynthesis.

5.2.1. Reducing polysaccharide biosynthesis

Seeds are generally covered with seed coats. Members of the Brassica family, including Arabidopsis, rapeseed, and so on, accumulate large amounts of mucilage consisting of pectin [55, 56]. Therefore, a considerable volume of the carbon source is consumed by mucilage production. Shen et al. reported that knockout of *GLABRA2*, which encodes a WRKY transcription factor controlling epidermal development, increases seed oil content in Arabidopsis [57]. Subsequent research demonstrated that increasing oil contents by suppressing *GLABRA2* expression reduces mucilage biosynthesis [58]. These findings indicate that reducing mucilage biosynthesis causes the carbon source to be directed into TAG biosynthesis. This transfer of the carbon source via reduction of polysaccharide levels has been verified in leaves. Starch comprises one of the main storage reserves in leaves. On the other hand, TAG biosynthesis activity is quite low in leaves. Therefore, to accumulate TAG in leaves, it is essential to both suppress starch biosynthesis and activate TAG biosynthesis. Sanjaya et al. found that overexpressing *WRI1* while suppressing the production of a small subunit of adenosine diphospho (ADP)-glucose pyrophosphorylase, a key enzyme for starch biosynthesis, significantly increases TAG contents in leaves [59]. These results indicate that reducing polysaccharide biosynthesis leads to the funneling of the carbon source into TAG biosynthesis, representing a possible novel strategy for increasing seed TAG contents in oil crops.

5.2.2. Reducing seed storage protein biosynthesis

As described above, oil seeds accumulate large amounts of seed storage proteins in addition to TAG. Crop breeders have reported a negative correlation between TAG and protein contents, especially in soybean [60, 61], suggesting that reducing the levels of seed storage proteins increases TAG contents in seeds. However, in Arabidopsis, knockout of genes encoding major seed storage proteins has little effect on TAG contents in seeds, although the protein content is reduced [54, 62]. These results may be due to the time lag between TAG and protein biosynthesis (**Figure 5**): little of the surplus carbon source derived from the suppression of protein biosynthesis is utilized for TAG biosynthesis because the TAG biosynthesis activity is quite low when protein synthesis is active in the late phase of seed development [54]. Therefore, simultaneously overexpressing *WRI1* and reducing protein synthesis greatly increases seed TAG contents due to effective utilization of the surplus carbon source [54]. This finding indicates that reducing protein synthesis indeed provides the surplus carbon source required for TAG production which, when combined with the simultaneous activation of TAG biosynthesis, leads to increased TAG production. This finding also demonstrates that the combination of these functional strategies has an additive effect on seed TAG content (**Figure 6**) [54, 63]. Breeding lines with reduced SSP content have been established in several crops [64, 65]. Introduction of *WRI1* into these breeding lines represents a potentially important strategy for high-oil seed production.

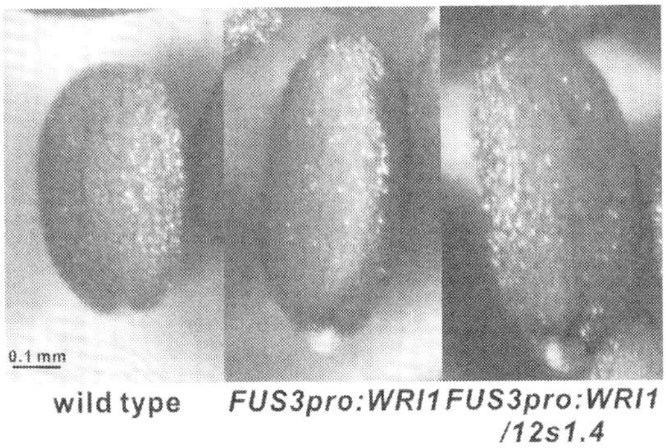

Figure 6. During the enlargement of seeds, the activation of TAG biosynthesis and the reduction of protein synthesis occur simultaneously. *FUS3pro:WRI1*; seeds from transgenic plants harboring *WRI1* under the control of the *FUS3* promoter in wild-type plants (Col-0): *WRI1* is expressed during the window phase. *FUS3pro:WRI1/12s1.4*; seeds from transgenic plants harboring *WRI1* under the control of the *FUS3* promoter in a double knockout mutant of *12S1* and *12S4*, encoding major seed storage proteins in *Arabidopsis thaliana*.

6. Conclusion

Molecular genetic analysis of Arabidopsis has led to the incredibly rapid elucidation of the mechanisms underlying the metabolism and regulation of seed storage reserves. The results from these basic studies provide powerful clues to help solve important issues in crop breeding. Optimizing the strategies developed based on the results of basic studies for use in crops will lead to crucial innovations for improving crop yields.

Author details

Masatake Kanai[1,2], Shoji Mano[1,2,3], Makoto Hayashi[4] and Mikio Nishimura[1*]

*Address all correspondence to: mikosome@nibb.ac.jp

1 Department of Cell Biology, National Institute for Basic Biology, Okazaki, Japan

2 Laboratory of Biological Diversity, Department of Evolutionary Biology and Biodiversity, National Institute for Basic Biology, Okazaki, Japan

3 Department of Basic Biology, SOKENDAI (The Graduate University for Advanced Studies), Okazaki, Japan

4 Department of Bioscience, Nagahama Institute of Bio-Science and Technology, Nagahama, Japan

References

[1] Hills MJ: Control of storage-product synthesis in seeds. Current Opinion in Plant Biology. 2004;7:302–308. DOI: 10.1016/j.pbi.2004.03.003.

[2] Santos-Mendoza M, Dubreucq B, Baud S, Parcy F, Caboche M, Lepiniec L: Deciphering gene regulatory networks that control seed development and maturation in Arabidopsis. The Plant Journal. 2008;54:608–620. DOI: 10.1111/j.1365-313X.2008.03461.x.

[3] Smith AM, Zeeman SC, Smith SM: Starch degradation. Annual Review of Plant Biology. 2005;56:73–98. DOI: 10.1146/annurev.arplant.56.032604.144257.

[4] Zeeman SC, Kossmann J, Smith AM: Starch: Its metabolism, evolution, and biotechnological modification in plants. Annual Review of Plant Biology, Vol 61. 2010;61:209–234. DOI: 10.1146/annurev-arplant-042809-112301.

[5] Bourgis F, Kilaru A, Cao X, Ngando-Ebongue GF, Drira N, Ohlrogge JB, Arondel V: Comparative transcriptome and metabolite analysis of oil palm and date palm mesocarp that differ dramatically in carbon partitioning. Proceedings of the National Academy of Sciences of the United States of America. 2011;108:18186–18186. DOI: 10.1073/pnas.1115243108.

[6] Dussert S, Guerin C, Andersson M, Joet T, Tranbarger TJ, Pizot M, Sarah G, Omore A, Durand-Gasselin T, Morcillo F: Comparative transcriptome analysis of three oil palm fruit and seed tissues that differ in oil content and fatty acid composition. Plant Physiology. 2013;162:1337–1358. DOI: 10.1104/pp.113.220525.

[7] Wang XM: Plant phospholipases. Annual Review of Plant Physiology and Plant Molecular Biology. 2001;52:211–231. DOI: 10.1146/annurev.arplant.52.1.211.

[8] Samuels L, Kunst L, Jetter R: Sealing plant surfaces: Cuticular wax formation by epidermal cells. Annual Review of Plant Biology. 2008;59:683–707. DOI: 10.1146/annurev.arplant.59.103006.093219.

[9] Dudareva N, Negre F, Nagegowda DA, Orlova I: Plant volatiles: Recent advances and future perspectives. Critical Reviews in Plant Sciences. 2006;25:417–440. DOI: 10.1080/07352680600899973.

[10] Schwab W, Davidovich-Rikanati R, Lewinsohn E: Biosynthesis of plant-derived flavor compounds. The Plant Journal. 2008;54:712–732. DOI: 10.1111/j.1365-313X.2008.03446.x.

[11] Weber H: Fatty acid-derived signals in plants. Trends in Plant Science. 2002;7:217–224. DOI:10.1016/S1360-1385(02)02250–1.

[12] Baud S, Lepiniec L: Physiological and developmental regulation of seed oil production. Progress in Lipid Research. 2010;49:235–249. DOI: 10.1016/j.plipres.2010.01.001.

[13] Durrett TP, Benning C, Ohlrogge J: Plant triacylglycerols as feedstocks for the production of biofuels. The Plant Journal. 2008;54:593–607. DOI: 10.1111/j.1365-313X.2008.03442.x.

[14] Ohlrogge JB, Jaworski JG: Regulation of fatty acid synthesis. Annual Review of Plant Physiology and Plant Molecular Biology. 1997;48:109–136. DOI: DOI 10.1146/annurev.arplant.48.1.109.

[15] Bates PD, Browse J: The significance of different diacylgycerol synthesis pathways on plant oil composition and bioengineering. Frontiers in Plant Science. 2012;3 DOI: 10.3389/fpls.2012.00147.

[16] Bates PD, Stymne S, Ohlrogge J: Biochemical pathways in seed oil synthesis. Current Opinion in Plant Biology. 2013;16:358–364. DOI: 10.1016/j.pbi.2013.02.015.

[17] Graham IA: Seed storage oil mobilization. Annual Review of Plant Biology. 2008;59:115–142. DOI: 10.1146/annurev.arplant.59.032607.092938.

[18] Napier JA, Beaudoin F, Tatham AS, Alexander LG, Shewry PR: The seed oleosins: Structure, properties and biological role. Advances in Botanical Research, Vol 35. 2001;35:111–138. DOI: 10.1016/S0065-2296(01)35005-X.

[19] Shimada TL, Hara-Nishimura I: Leaf oil bodies are subcellular factories producing antifungal oxylipins. Current Opinion in Plant Biology. 2015;25:145–150. DOI: 10.1016/j.pbi.2015.05.019.

[20] Napier JA, Stobart AK, Shewry PR: The structure and biogenesis of plant oil bodies: The role of the er membrane and the oleosin class of proteins. Plant Molecular Biology. 1996;31:945–956. DOI: 10.1007/Bf00040714.

[21] Tzen JTC, Cao YZ, Laurent P, Ratnayake C, Huang AHC: Lipids, proteins, and structure of seed oil bodies from diverse species. Plant Physiology. 1993;101:267–276.

[22] Siloto RMP, Findlay K, Lopez-Villalobos A, Yeung EC, Nykiforuk CL, Moloney MM: The accumulation of oleosins determines the size of seed oilbodies in Arabidopsis. The Plant Cell. 2006;18:1961–1974. DOI: 10.1105/tpc.106.041269.

[23] Skoric D: Achievements and future-directions of sunflower breeding. Field Crops Research. 1992;30:231–270. DOI: Doi 10.1016/0378-4290(92)90003-R.

[24] Zheng P, Allen WB, Roesler K, Williams ME, Zhang S, Li J, Glassman K, Ranch J, Nubel D, Solawetz W, Bhattramakki D, Llaca V, Deschamps S, Zhong GY, Tarczynski MC, Shen B: A phenylalanine in dgat is a key determinant of oil content and composition in maize. Nature Genetics. 2008;40:367–372. DOI: 10.1038/ng.85.

[25] Li HW, Zhao TJ, Wang YF, Yu DY, Chen SY, Zhou RB, Gai JY: Genetic structure composed of additive QTL, epistatic QTL pairs and collective unmapped minor QTL conferring oil content and fatty acid components of soybeans. Euphytica. 2011;182:117–132. DOI: 10.1007/s10681-011-0524-9.

[26] Hwang EY, Song QJ, Jia GF, Specht JE, Hyten DL, Costa J, Cregan PB: A genome-wide association study of seed protein and oil content in soybean. Bmc Genomics. 2014;15 DOI: 10.1186/1471-2164-15-1.

[27] Burns MJ, Barnes SR, Bowman JG, Clarke MHE, Werner CP, Kearsey MJ: QTL analysis of an intervarietal set of substitution lines in *Brassica napus*: seed oil content and fatty acid composition. Heredity. 2003;90:39–48. DOI: 10.1038/sj.hdy.6800176.

[28] Jiang CC, Shi JQ, Li RY, Long Y, Wang H, Li DR, Zhao JY, Meng JL: Quantitative trait loci that control the oil content variation of rapeseed (*Brassica napus L.*). Theoretical and Applied Genetics. 2014;127:957–968. DOI: 10.1007/s00122-014-2271-5.

[29] McGlew K, Shaw V, Zhang M, Kim RJ, Yang W, Shorrosh B, Suh MC, Ohlrogge J: An annotated database of Arabidopsis mutants of acyl lipid metabolism. Plant Cell Reports. 2015;34:519–32. DOI: 10.1007/s00299-014-1710-8.

[30] Li-Beisson Y, Shorrosh B, Beisson F, Andersson MX, Arondel V, Bates PD, Baud S, Bird D, Debono A, Durrett TP, Franke RB, Graham IA, Katayama K, Kelly AA, Larson T, Markham JE, Miquel M, Molina I, Nishida I, Rowland O, Samuels L, Schmid KM, Wada H, Welti R, Xu C, Zallot R, Ohlrogge J: Acyl-lipid metabolism. The Arabidopsis book / American Society of Plant Biologists. 2013;11:e0161. DOI: 10.1199/tab.0161.

[31] Jako C, Kumar A, Wei YD, Zou JT, Barton DL, Giblin EM, Covello PS, Taylor DC: Seed-specific over-expression of an Arabidopsis cDNA encoding a diacylglycerol acyltransferase enhances seed oil content and seed weight. Plant Physiology. 2001;126:861–874. DOI: 10.1104/Pp.126.2.861.

[32] Vigeolas H, Waldeck P, Zank T, Geigenberger P: Increasing seed oil content in oil-seed rape (*Brassica napus L.*) by over-expression of a yeast glycerol-3-phosphate dehydrogenase under the control of a seed-specific promoter. Plant Biotechnology Journal. 2007;5:431–441. DOI: 10.1111/j.1467-7652.2007.00252.x.

[33] Lardizabal K, Effertz R, Levering C, Mai J, Pedroso MC, Jury T, Aasen E, Gruys K, Bennett K: Expression of Umbelopsis ramanniana DGAT2A in seed increases oil in soybean. Plant Physiology. 2008;148:89–96. DOI: 10.1104/pp.108.123042.

[34] Andrianov V, Borisjuk N, Pogrebnyak N, Brinker A, Dixon J, Spitsin S, Flynn J, Matyszczuk P, Andryszak K, Laurelli M, Golovkin M, Koprowski H: Tobacco as a production platform for biofuel: Overexpression of *Arabidopsis* DGAT and LEC2 genes increases accumulation and shifts the composition of lipids in green biomass. Plant Biotechnology Journal. 2010;8:277–287. DOI: 10.1111/j.1467-7652.2009.00458.x.

[35] Wang ZK, Huang WJ, Chang JM, Sebastian A, Li YG, Li HY, Wu XX, Zhang BB, Meng FL, Li WB: Overexpression of SiDGAT1, a gene encoding acyl-CoA:diacylglycerol acyltransferase from Sesamum indicum L. Increases oil content in transgenic Arabidopsis and soybean. Plant Cell Tissue and Organ Culture. 2014;119:399–410. DOI: 10.1007/s11240-014-0543-z.

[36] Taylor DC, Zhang Y, Kumar A, Francis T, Giblin EM, Barton DL, Ferrie JR, Laroche A, Shah S, Zhu W, Snyder CL, Hall L, Rakow G, Harwood JL, Weselake RJ: Molecular modification of triacylglycerol accumulation by over-expression of DGAT1 to produce canola with increased seed oil content under field conditions. Botany-Botanique. 2009;87:533–543. DOI: 10.1139/B08-101.

[37] Weselake RJ, Shah S, Tang MG, Quant PA, Snyder CL, Furukawa-Stoffer TL, Zhu WM, Taylor DC, Zou JT, Kumar A, Hall L, Laroche A, Rakow G, Raney P, Moloney MM, Harwood JL: Metabolic control analysis is helpful for informed genetic manipulation of oilseed rape (Brassica napus) to increase seed oil content. Journal of Experimental Botany. 2008;59:3543–3549. DOI: 10.1093/jxb/ern206.

[38] Mano S, Nishimura M: Plant peroxisomes. Plant Hormones. 2005;72:111–154. DOI: 10.1016/S0083-6729(05)72004-5.

[39] Theodoulou FL, Eastmond PJ: Seed storage oil catabolism: A story of give and take. Current Opinion in Plant Biology. 2012;15:322–328. DOI: 10.1016/j.pbi.2012.03.017.

[40] Carrera E, Holman T, Medhurst A, Peer W, Schmuths H, Footitt S, Theodoulou FL, Holdsworth MJ: Gene expression profiling reveals defined functions of the ATP-binding cassette transporter comatose late in phase ii of germination. Plant Physiology. 2007;143:1669–1679. DOI: 10.1104/pp.107.096057.

[41] Kamada T, Nito K, Hayashi H, Mano S, Hayashi M, Nishimura M: Functional differentiation of peroxisomes revealed by expression profiles of peroxisomal genes in Sesamum indicum L. Plant and Cell Physiology. 2003;44:1275–1289. DOI: 10.1093/Pcp/Pcg173.

[42] Kanai M, Nishimura M, Hayashi M: A peroxisomal ABC transporter promotes seed germination by inducing pectin degradation under the control of ABI5. The Plant Journal. 2010;62:936–947. DOI: 10.1111/j.1365-313X.2010.04205.x.

[43] Turley RB, Trelease RN: Development and regulation of three glyoxysomal enzymes during cotton seed maturation and growth. Plant Molecular Biology. 1990;14:137-46.

[44] Comai L, Dietrich RA, Maslyar DJ, Baden CS, Harada JJ: Coordinate expression of transcriptionally regulated isocitrate lyase and malate synthase genes in *Brassica-napus L*. The Plant Cell. 1989;1:293–300. DOI: 10.1105/Tpc.1.3.293.

[45] Chia TYP, Pike MJ, Rawsthorne S: Storage oil breakdown during embryo development of Brassica napus (L.). Journal of Experimental Botany. 2005;56:1285–1296. DOI: 10.1093/jxb/eri129.

[46] Chen H, Wang FW, Dong YY, Wang N, Sun YP, Li XY, Liu L, Fan XD, Yin HL, Jing YY, Zhang XY, Li YL, Chen G, Li HY: Sequence mining and transcript profiling to explore differentially expressed genes associated with lipid biosynthesis during soybean seed development. Bmc Plant Biology. 2012;12 DOI: 10.1186/1471-2229-12-122.

[47] Kelly AA, Shaw E, Powers SJ, Kurup S, Eastmond PJ: Suppression of the *SUGAR-DEPENDENT1* triacylglycerol lipase family during seed development enhances oil

yield in oilseed rape (*Brassica napus L.*). Plant Biotechnology Journal. 2013;11:355–361. DOI: 10.1111/pbi.12021.

[48] Eastmond PJ: SUGAR-DEPENDENT1 encodes a patatin domain triacylglycerol lipase that initiates storage oil breakdown in germinating Arabidopsis seeds. The Plant Cell. 2006;18:665–675. DOI: 10.1105/tpc.105.040543.

[49] Kim MJ, Yang SW, Mao HZ, Veena SP, Yin JL, Chua NH: Gene silencing of *Sugar-dependent 1 (JcSDP1)*, encoding a patatin-domain triacylglycerol lipase, enhances seed oil accumulation in Jatropha curcas. Biotechnology for Biofuels. 2014;7 DOI: 10.1186/1754-6834-7-36.

[50] Maeo K, Tokuda T, Ayame A, Mitsui N, Kawai T, Tsukagoshi H, Ishiguro S, Nakamura K: An AP2-type transcription factor, WRINKLED1, of Arabidopsis thaliana binds to the AW-box sequence conserved among proximal upstream regions of genes involved in fatty acid synthesis. The Plant Journal. 2009;60:476–487. DOI: 10.1111/j.1365-313X.2009.03967.x.

[51] Baud S, Wuilleme S, To A, Rochat C, Lepiniec L: Role of WRINKLED1 in the transcriptional regulation of glycolytic and fatty acid biosynthetic genes in Arabidopsis. The Plant Journal. 2009;60:933–947. DOI: 10.1111/j.1365-313X.2009.04011.x.

[52] Baud S, Mendoza MS, To A, Harscoet E, Lepiniec L, Dubreucq B: WRINKLED1 specifies the regulatory action of LEAFY COTYLEDON2 towards fatty acid metabolism during seed maturation in Arabidopsis. The Plant Journal. 2007;50:825–838. DOI: 10.1111/j.1365-313X.2007.03092.x.

[53] Cernac A, Benning C: *WRINKLED1* encodes an AP2/EREB domain protein involved in the control of storage compound biosynthesis in Arabidopsis. The Plant Journal. 2004;40:575–585. DOI: 10.1111/j.1365-313X.2004.02235.x.

[54] Kanai M, Mano S, Kondo M, Hayashi M, Nishimura M: Extension of oil biosynthesis during the mid-phase of seed development enhances oil content in *Arabidopsis seeds*. Plant Biotechnology Journal. 2016;14:1241–1250. DOI: 10.1111/pbi.12489.

[55] Western TL, Skinner DJ, Haughn GW: Differentiation of mucilage secretory cells of the Arabidopsis seed coat. Plant Physiology. 2000;122:345-355. DOI:10.1104/Pp.122.2.345.

[56] Haughn GW, Western TL: *Arabidopsis*seed coat mucilage is a specialized cell wall that can be used as a model for genetic analysis of plant cell wall structure and function. Frontiers in Plant Science. 2012;3 DOI: 10.3389/Fpls.2012.00064.

[57] Shen B, Sinkevicius KW, Selinger DA, Tarczynski MC: The Homeobox gene *GLABRA2 affects seed oil content in Arabidopsis*. Plant Molecular Biology. 2006;60:377–387. DOI: 10.1007/s11103-005-4110-1.

[58] Shi L, Katavic V, Yu YY, Kunst L, Haughn G: Arabidopsis *glabra2* mutant seeds deficient in mucilage biosynthesis produce more oil. The Plant Journal. 2012;69:37–46. DOI: 10.1111/j.1365-313X.2011.04768.x.

[59] Sanjaya, Durrett TP, Weise SE, Benning C: Increasing the energy density of vegetative tissues by diverting carbon from starch to oil biosynthesis in transgenic Arabidopsis. Plant Biotechnology Journal. 2011;9:874–883. DOI: 10.1111/j.1467-7652.2011.00599.x.

[60] Eskandari M, Cober ER, Rajcan I: Genetic control of soybean seed oil: Ii. QTL and genes that increase oil concentration without decreasing protein or with increased seed yield. Theoretical and Applied Genetics. 2013;126:1677–1687. DOI: 10.1007/s00122-013-2083-z.

[61] Mao TT, Jiang ZF, Han YP, Teng WL, Zhao X, Li WB: Identification of quantitative trait loci underlying seed protein and oil contents of soybean across multi-genetic backgrounds and environments. Plant Breeding. 2013;132:630–641. DOI: 10.1111/pbr.12091.

[62] Lin YM, Pajak A, Marsolais F, McCourt P, Riggs CD: Characterization of a cruciferin deficient mutant of Arabidopsis and its utility for overexpression of foreign proteins in plants. Plos One. 2013;8 DOI: 10.1371/journal.pone.0064980.

[63] van Erp H, Kelly AA, Menard G, Eastmond PJ: Multigene engineering of boosts seed oil content in Arabidopsis. Plant Physiology. 2014;165:30-6. DOI: 10.1104/pp.114.236430.

[64] Schmidt MA, Barbazuk WB, Sandford M, May G, Song ZH, Zhou WX, Nikolau BJ, Herman EM: Silencing of soybean seed storage proteins results in a rebalanced protein composition preserving seed protein content without major collateral changes in the metabolome and transcriptome. Plant Physiology. 2011;156:330–345. DOI: 10.1104/pp.111.173807.

[65] Kawakatsu T, Hirose S, Yasuda H, Takaiwa F: Reducing rice seed storage protein accumulation leads to changes in nutrient quality and storage organelle formation. Plant Physiology. 2010;154:1842–1854. DOI: 10.1104/pp.110.164343.

Chapter 7

The Dynamics of Plant Cell Wall *In Muro* Modifications and its Physiological Implications on Seed Germination

Ximena Gómez-Maqueo and
Alicia Gamboa-deBuen

Additional information is available at the end of the chapter

http://dx.doi.org/10.5772/64085

Abstract

Seed germination is a complex process in which the embryo, enclosed within the surrounding tissues, must quickly switch from a maturation program to a germination-driven developmental process that will prepare the embryo for seedling growth and establishment. The germination process initiates with water uptake by the dry seed and culminates, usually, with the radicle protrusion. The radicle emergence from the seed is a highly regulated process that involves discrete and coordinated changes in plant cell wall extensibility and rearrangements of its components, among other processes. In this chapter we will review current knowledge of the physiological process of controlled cell separation and expansion, which give the primary cell wall its plastic properties by "loosening" of the main components of the cell wall during seed germination. We will focus on the physiological importance of primary cell wall constitution and modification by the activity *in muro* of a broad variety of cell wall-modifying enzymes that include hydrolases and transglycosylases, as well as non-enzymatic processes such as expansin-mediated loosening during seed germination.

Keywords: cell wall modification, primary cell wall, seed germination

1. Introduction

Seeds constitute a critical stage in the life cycle of embryophytes. In this stage, the plant embryo remains in a quiescent state until the proper conditions of temperature, water availability, and, in some species, light are met in order for the processes of germination and seedling establishment to occur [1, 2]. The mature seed contains the embryo, which is surrounded by the seed coat (testa) that is derived from the maternal tissues and in some species by one or more layers of storage tissue (endosperm) [2]. Seeds can function as resistance structures. Several mechanisms have evolved, in tight relation with the environment, to ensure the survival of the quiescent embryo [3]. Part of these mechanisms includes the modification of the structure and composition of plant cell walls.

One characteristic feature of plant cells is that they are enclosed in a polysaccharide and protein matrix, denominated as cell wall [4]. Plant cells can have two different types of wall. Primary walls, produced during cytokinesis, are flexible structures that regulate cell growth and shape. The secondary walls are deposited after the cell has achieved its final size and shape, by the inclusion of lignin and other phenolic compounds, thus making the cell wall rigid and usually impermeable. Cell walls have several functions that include the regulation of cell-cell adhesion and abscission, apoplastic transport, mechanical support and maintenance of turgor pressure, and defense against pathogens [2, 5]. In seeds, cell walls are modified in order to generate hard, and in some cases impermeable, coats that protect the embryo from the environmental conditions. Also, seed cell walls can store energy that can be mobilized to feed embryo growth and development. Finally, cell walls regulate the timing of seed germination by fine-tuning the processes of matrix polysaccharide loosening/breakage, as well as the integration of environmental cues with the hormonal and physiological status of the embryo [4, 6]. In this chapter we will focus only on primary cell walls and their importance on seed germination.

2. Seed germination

Seed germination is a physiological process initiated with water uptake and culminating with the emergence of the embryo through its protective tissues, which might include the testa, endosperm, perisperm, or pericarp [2]. The testa and the endosperm rupture must be coordinated with environmental seasonality to facilitate germination in the most favorable conditions [1, 6]. Several mechanisms have evolved to ensure proper synchronization of germination with environmental cues; among these is the interplay of hormonal signaling pathways via abscisic acid (ABA), gibberellins (GA), ethylene, and jasmonates [7–10]. These hormones exert their regulation on germination through different pathways including cell wall remodeling [7, 11].

In the classical model of seed germination described by Bewley et al. [12], the process of germination is divided into three phases, distinguished by the rate of water absorption by the seed tissues. The phase I, or imbibition phase, is characterized by a rapid water uptake rate driven by the difference in water potential between the seed and the environment. In this phase also the reactivation of primary metabolism and DNA repair pathways starts. Next, in phase II or activation phase, the imbibition rate decreases, water content remains stable, and major changes in the metabolic pathways and activation of other cellular processes take place. In this phase the integration of environmental cues with the internal status of the seed that will determine whether or not the seed will enter into the next phase occurs. Finally, in phase III there is another rapid water uptake driven by radicle protrusion and is mainly related to seedling growth. Germination is completed once the radicle has emerged at the onset of phase

III. This triphasic model of imbibition can be applied to all seeds analyzed thus far [12, 13]. The imbibition time needed for completion of germination is highly variable among species and even within seed lots, and it depends on several factors like seed history and environmental conditions experienced by the mother plant at the moment of seed dispersion and during the after-ripening period [12, 14, 15].

It is now generally accepted that radicle protrusion occurs by two nonexclusive processes [2, 13]. The first process involves a decrease in the mechanical resistance of the enclosing tissues, especially in the micropylar region of the testa and endosperm [2, 10]. The second process deals with an increasing growth potential of the embryo, driven by turgor pressure and cellular expansion in the embryonic axis [2, 13]. Most of the knowledge generated about the regulation of radicle protrusion comes from endospermic seeds, where testa and endosperm rupture can occur in two easily distinguishable stages (*Arabidopsis thaliana –Arabidopsis–, Chenopodium album, Lepidium* sp., *Nicotiana* sp., to mention a few) [2].

In recent years, with the advent of whole genome/transcriptome analysis, it has been possible to study the process of germination with high spatial-temporal resolution. Transcriptomic analysis allows a comprehensive view of seed germination by dissecting "early" or "late" germination processes, the first being the initial response to water and the second corresponding to the interval from the imbibed seed to the radicle protrusion [14, 15]. Also, in endospermic seeds, an important landmark is the distinction between the processes that occur prior to testa rupture and after it that leads to endosperm rupture [8, 16–18].

Several studies demonstrate that the main transcripts, enzymes, and other proteins accumulated in dry seeds participate in primary metabolism, starch and storage protein mobilization, reactive oxygen species (ROS) scavenging, and cell wall synthesis [14, 15]. Aside from providing building blocks to sustain protein production and cell growth, the reactivation of primary metabolism in the early stages of seed germination plays a major role in the generation of the proper redox state to promote the activity of different enzymes and produce energy to support processes essential for radicle protrusion [14, 19].

In *Arabidopsis*, the seed development and maturation programs are regulated by the LAFL transcription factor network (LEAFY COTYLEDON 1 (LEC1) and LEC1-LIKE (L1L), ABA INSENSITIVE 3 (ABI3), FUSCA 3 (FUS3), and LEC2), which activates other downstream transcription factor networks in concerted action of hormone, sugar, and light signalization pathways. Some target genes are involved in ABA, GA, ethylene, brassinoesteroids (BR), auxin, jasmonic acid (JA), and cytokinin (CK) signalization pathways [20]. The ABA signalization pathway participates in the regulatory networks of seed maturation, reserve accumulation, and desiccation tolerance acquisition [21]. GA blocks the LAFL and ABA networks during germination. The degradation of transcripts and enzymes related to seed maturation, which accumulated in the dry seed, has been described to occur in the first 6–12 h of seed imbibition in *Arabidopsis* [22] and within the first 24 h in *rice* and *barley* [14, 23].

Gibberellins play a major role in promoting a myriad of developmental programs, and its antagonistic role in ABA-mediated block of germination has been described [24]. GA stimulates seed germination by enhancing embryo growth; embryos of *Arabidopsis* GA-deficient mutant seeds exhibit reduced growth rate phenotypes [25]. Also, GA enhances seed germination by overcoming the mechanical restraint to radicle protrusion of the surrounding tissues. In *Solanum lycopersicum* (tomato), GA-deficient embryos (unable to germinate unless incubated with exogenous GA) can grow into dwarf plants when the testa and the endosperm were removed mechanically [26]. This role of GAs in stimulating germination can be linked to the upregulation of several cell wall-modifying proteins (CWMPs) detected in whole-seed *Arabidopsis* transcriptomes of *ga1-3* mutants treated with GA_4 [24]. Jacobsen and Pressman [27] suggested that the embryo of celery (*Apium graveolens*) seeds does not secrete CWMPs but rather promote the activity of GA-inducible CWMPs in the endosperm. The depletion of the endosperm in this species generates a space where the embryo cells can expand and eventually penetrate the micropylar endosperm.

An overrepresentation analysis (ORA) of gene ontologies showed that transcription regulation is enriched in both the endosperm and the embryo transcriptomes of *Arabidopsis* seeds. The ORA analysis also showed that in the endosperm, the main biological processes are associated with cell wall metabolism, cell death, response to biotic stimulus, and defense and response to ABA. The main biological processes in the embryo include phosphate metabolic process, protein amino acid phosphorylation, hormone metabolic process (particularly auxin synthesis and transport), cell division and cell cycle, post-germination regulation of growth and organ development, and signaling [17].

3. Cell wall structure and composition

Plant cell walls are complex and highly dynamic structures composed of a variety of polysaccharides, proteins, and aliphatic or aromatic compounds [28, 29]. They are continually being modified throughout development and in response to environmental stimuli [30, 31]. Primary cell walls of flowering plants can be classified in two main groups depending on its general architecture and composition, as well as their biosynthetic processes [32, 33]. Type I cell walls are the most common, present in dicotyledonous and the non-commelinoid monocotyledonous plants (a more basal group of aroids, alismatids, and lilioids). Type II cell walls are found only in the commelinoid monocots that include the Poales (members of the families Poaceae, Bromeliaceae, and Cyperaceae) [32, 33].

3.1. Primary cell wall polysaccharides

Primary cell wall polysaccharides constitute the majority of the wall dry mass in land plants and can be grouped in three main classes: cellulose, hemicelluloses, and pectins [30]. Cellulose is a linear 1,4-β-D-glucan that assembles into partially crystalline microfibrils, each of which contains about 36 parallel polysaccharide chains [34]. Cellulose is synthesized *in muro* by the cellulose synthase complex (CSC), embedded within the plasma membrane and formed by 6 rosette subunits that contain 6 cellulose synthase proteins (CESA) [35]. Aside from cellulose, all other cell wall polysaccharides are synthesized and processed for wall targeting in the trans-Golgi system [5].

Hemicelluloses are polymers whose backbones consist of β-glucose, β-xylose, or β-mannose, with short side chains. In all vascular plants with type I walls, the most common hemicelluloses are xyloglucans (XyG), whereas type II cell walls contain less XyG, being the most abundant glucuronoarabinoxylans (GAX) and β1,3:β1,4 mixed glucans [33]. Hemicellulose chains adhere to cellulose microfibrils, in a rope-like manner, to restrain cell expansion [30, 34]. Also, in type I cell walls, hemicelluloses bind and cross-link with pectin and form the hydrated matrix [30].

The group generally known as pectins comprises over 30% of the cell wall total mass in dicots [31, 36]. Pectins are acidic heteropolymers that form a hydrated gel, in which cellulose and other molecules are embedded in the plant cell wall. Their main defining feature is 1,4-linked α-D-galacturonic acid residues (GalA). Pectins interact covalently and non-covalently with other pectin molecules or with hemicellulose xyloglucan or arabinogalactans [31]. Several studies support the hypothesis that the three major pectin classes, homoglacturonan (HG), rhamnogalacturonan I, and rhamnogalacturonan II, are covalently linked in the cell wall [29, 37], forming a hydrophilic macromolecular network. Pectin is deposited on the cell wall matrix in a highly methylesterified form [28]. The methyl group is removed by pectin methylesterases (PMEs) in muro, providing an anionically charged matrix and changing the mechanic properties of the cell wall. Increasing evidence shows that the regulation of the degree of methylesterification of the pectic matrix plays a fundamental role in plant growth, development, morphogenesis, cell-cell adhesion, cell expansion, seed hydration, and seed germination [5, 16, 36, 38].

Other polysaccharides present in primary cell walls of various species are the mannans, arabinoxylans, and arabinogalactans. Mannans are formed by mannosyl residues linked by β-1,4-glycosidic linkages. This mannosyl backbone can contain glucose residues (glucomannans) or be further substituted by single galactose residues with α-1,6-linkages (galactomannans). Arabinoxylan consists of a (1,4)-linked β-D-xylan backbone decorated with arabinose branches. Other residues, such as glucuronic acid and ferulic acid esters (FAE), are also attached in arabinoxylans that are particularly abundant in cereal grasses. Arabinogalactan and storage xyloglucans are used as reserves in cotyledons. The basic structure of storage xyloglucans differs from the primary wall xyloglucans in that it is not fucosylated [39].

3.2. Cell wall proteins

Primary cell walls are mainly constituted by polysaccharides; however, proteins account for about 10% of the total dry mass of the wall [40]. Proteins that contain a secretion signal peptide, which targets them to the secretory pathway and in most cases is excised to allow activation or proper protein function, are commonly referred as classical cell wall proteins [40–42]. In *Arabidopsis*, about 17% (~5000 genes) of the genome encodes for proteins targeted to the secretory pathway, and of this, about 1000–2000 genes could be cell wall proteins (CWPs). Cell wall proteins have several functions as structural, enzymatic, and defense and have been grouped in functional categories by different authors. The proteins with a structural function or those acting on polysaccharides are the two main functional categories [30, 42]. These proteins are expressed in a tissue-specific and process-specific manner, contributing to the regulation of cell wall stabilization and rigidity [41].

3.2.1. Structural proteins

Structural proteins are usually classified by the predominant amino acids in their sequence, although some of them can belong to more than one category. The most common include the hydroxyproline-rich glycoproteins (HRGPs), the glycine-rich proteins (GRPs), the proline-rich proteins (PRPs), and the arabinogalactan proteins (AGPs). These proteins vary greatly in abundance within plant species, cell tissues, and environmental conditions [4]. Arabinogalactan proteins, that are widely distributed among plant families and comprise about 2–10% of the total protein in the wall, are highly glycosylated. Also, AGPs are rich in hydroxyproline, serine, alanine, threonine, and glycine, and resistant to proteolysis in their native state [41]. Extensins are a family of HRGPs particularly abundant in dicots that have been involved in modification of wall extensibility in elongating tissues [43].

3.2.2. Proteins acting on polysaccharides

Within the CWPs acting on polysaccharides, there are a broad variety of activities. For instance, the group of glycosyl hydrolases (GHs) include glycosidases (β-glucosidase, β-galactosidase, β-xylosidase, and α-xylosidase, and exo-polygalacturonases) and glycanases like β-mannanase, β-xylanase, (1→4)-β-glucanase "cellulase", endo-polygalacturonases, and xyloglucan endo-hydrolase (XEH). The combined activity of these kinds of enzymes is theoretically capable of hydrolyzing most of the glycosidic bonds in the cell wall polysaccharides but do not imply that all enzymes are active at the same time or tissue. The glycosyltransferases (GTs) activity involves the formation of a glycosyl-enzyme complex that is attacked by an acceptor substrate (another oligo/polysaccharide). This activity allows the integration of recently secreted polysaccharides into the matrix and the grafting of polysaccharides already present in the wall matrix [44]. This category includes the xyloglucan endotransglycosylase (XET). Both XEH and XET proteins are commonly grouped within the xyloglucan endotransglucosylase/hydrolase family (XTH) due to some of their members (like β-xylanase) that can have both GH and GT activities [30].

Polysaccharide lyases (PLs) promote cell separation by calcium-dependent de-polymerization of wall polygalacturonides. Plant pectate lyases are a group of enzymes that catalyze the cleavage of de-methylesterified pectin. PL activity has being described in cell wall degradation that occurs during fruit ripening [45], and Penfield et al. [7] report 34 pectate lyases that were downregulated in the endosperm of *Arabidopsis* imbibed seeds after treatment with ABA.

Carbohydrate esterases (CEs) include two enzyme families that have activity over pectins, the PMEs and pectin acetylesterases (PAEs), and the family of xylan acetylesterases. These enzymes cleave methyl or acetyl groups from the HG or Xyl backbone of polysaccharides [30].

PMEs catalyze the reaction by which methylesters are cleaved from a HG chain, producing a free carboxyl group and the release of a proton and methanol [46]. Plant PMEs are mainly alkaline isoforms bound to the wall matrix, while some isoforms are neutral and easily solubilized or free apoplastic acidic isoforms. Alkaline isoforms seem to be the PMEs with most de-methylesterification activity, but the kinetics of PME activity is affected by the ionic composition of the matrix, thus influencing PME activity and mobility [47]. PME activity can lead to two different cell wall fates: the first one would be the formation of a rigid, stable structure by Ca^{2+} interaction with de-methylesterified GalA residues (>10) in the HG chains. The second fate of HG would be their degradation by polygalacturonases, where only small stretches or individual GalA residues are de-methylesterified, thus leading to a more relaxed matrix [28, 46]. Also, PME activity acidifies the cell wall; this acidification would allow expansin activity ("acid grow") [5]. PMEs are antagonistically regulated in the cell wall by proteinaceous PME inhibitors (PMEIs) meanwhile PGs by PG inhibitors (PGIPs) [28].

Expansins regulate cell wall loosening in a pH-dependent manner by disruption of the hydrogen bonds between xyloglucans and cellulose. Sequence analysis indicates that expansins contain an N-terminal domain slightly similar to the catalytic domain of the family-45 endoglucanases; however no catalytic activity has been reported.

In *Arabidopsis*, about 10% of the total CWPs described in cell wall proteomes from different tissues and plants correspond to gene families with domains of unknown function (DUF) [42]. Mewalal et al. [48] have pointed out the relevance of these DUF families on cell wall dynamics. In particular, the two plant-specific families DUF231 and DUF642 could be involved in pectin modification [49, 50].

3.3. Cell wall modification

The molecular modification of the wall network can result in the relaxation of wall stress or "wall loosening" by the controlled rearrangement of cellulose/matrix polymers, which involve sliding of a cross-link along a scaffold or the breakage of stress-bearing cross-links without substantial changes in wall dimensions. These rearrangements could include three processes: (a) the cleavage of the backbone of major matrix polymers, (b) the weakening of the non-covalent bonding between polysaccharides, and (c) the breakage of cross-links [5]. Following cell wall loosening, there are three main types of outcomes: cell expansion, cell separation, and wall stiffening. Cell wall enlargement occurs secondarily as a result of water uptake and the reduction of turgor pressure resulting from wall loosening [44].

Reactive oxygen species (ROS) like hydrogen peroxide (H_2O_2), hydroxyl radical (·OH), and superoxide radical (O_2^-) have been proposed to play a major role in germination by participating in defense against pathogens, signaling, and promotion of cell wall loosening [2, 9]. ROS can negatively affect germination by reacting with almost all macromolecules stored in the seed, causing oxidative damage and cleavage of polysaccharide chains in the cell walls [5, 9]. The participation of ROS in cell wall loosening and promotion of germination might be indirect, through the ethylene signaling pathways that involve ROS production and downstream activation of CWMPs [9]. Cosgrove [5] suggests the revision of ROS participation in the process of wall loosening, since in most studies reporting ROS-mediated extensibility comprises only a small fraction (about 1% extensibility) and the assays with higher ROS concentrations provoke wall breakage.

3.4. Role and regulation of cell wall enzymes and proteins during germination

The study of plant cell wall structure and physiology has achieved a major progress from the input of "-omics" technologies in the past two decades. These -omics technologies are able to capture the complexity of biological processes, like seed germination and cell wall modification, with high sensitivity and spatial-temporal resolution. A tissue-specific transcriptome analysis in *Arabidopsis* showed that both endosperm and embryo share gene expression patterns and biological processes during seed germination. About 10,800 transcripts (~84% of total genes expressed) are present in both endosperm and embryo. Endosperm-specific genes comprise about 415 genes that were highly expressed [17]. Of this gene set, 154 are cell wall-related genes, with most of them being expressed at the onset of testa rupture. Transcript abundance of several CWMPs shows a transient peak with a 6–24 h interval in tomato [51], *Arabidopsis* [17], and *Lepidium* seeds [16]. Although a change in transcript levels does not necessarily correlate to changes in protein abundance or enzymatic activity, it has been demonstrated that during *rice* germination, most cell wall-related transcripts, as well as the resulting metabolites from cell wall modification, accumulate about 12–24 h after imbibition (HAI, although some metabolites can be detected by 3 HAI) [23].

Cell wall modification can occur at five different stages during seed germination: (a) during the cellular expansion process triggered by rehydration of tissues, (b) at the onset of testa rupture, (c) during endosperm weakening and rupture, (d) during cellular expansion related to radicle elongation, and (e) during wall degradation and mobilization of stored reserves in both living and nonliving storage tissues.

3.4.1. *Rehydration-driven cellular expansion*

Seed imbibition is given by the difference in water potential between the seed and the environment. Nonviable seeds swell faster than viable seeds, as viable seeds develop turgor pressure that restricts further water uptake [2]. However, rapid imbibition can still occur and lead to solute leakage and damage of membranes. Gradual rehydration of seed tissues has been detected in legumes like peas and beans, where hydration starts in the tissues near to the micropyle. As water diffuses in the outermost tissues, a waterfront is formed between imbibed tissues and those about to be imbibed. The testa plays a significant role in modulating imbibition kinetics and the waterfront formation [2]. The seeds of mutants with altered testa structure or altered deposition of protecting substances (like flavonoids, cutin, suberin, and lignin) have increased permeability and lower longevity than the wild type [52]. Testa structure usually consists of several layers of highly compressed dead cells where protective substances are deposited during seed development and maturation. Plant cell walls of living cells can also function as an interface that modulates water intake by changing wall porosity, thus allowing a gradual swelling of all tissues. This regulation could be achieved by rapid changes in wall extensibility as the ones generated by expansins. In support of this view, in whole unstratified *Arabidopsis* seeds, the upregulation of *EXPA1, EXPA2, EXPA3, EXPA8, EXPA9, EXPA15,* and *EXPA20* transcripts from 0 to 12 h has been reported. This induction was evident in seeds imbibed in the light and during moist cold stratification at 4°C in the dark [11]. Also, in whole-seed transcriptomes, the expression of *AtXTH5, AtXTH6,* and *AtXTH33* transcripts within the first 6HAI and *AtXTH3* and *AtXTH*33 transcripts at 12HAI [24] was detected. The activity of XTHs from *chickpea* seeds has been detected in imbibed seeds from 1HAI until 24HAI, when the rate of radicle elongation slows down [53].

In many species from the Brassicaceae, Solanaceae, Linaceae, and Plantaginaceae, among others, the epidermis of the testa contains specialized cells that accumulate abundant pectins and heteroxylans, as well as some xyloglucans or arabinans during seed development. Upon imbibition these polysaccharides expand and burst out of the testa, generating a gel-like structure. This phenomenon, known as myxospermy, has been used as a model to study several hydrolases and PME activity [54, 55]. Although there is still uncertainty about the actual role of myxospermy, the proposed roles include regulating hydration, preventing desiccation, being an oxygen barrier, or allowing the seed to attach to the substrate and animals [54–56].

3.4.2. Testa rupture

In many plant species, testa rupture starts at the micropylar seed end. In tobacco (*Nicotiana tabacum*) seeds, the testa ruptures at the micropyle and follows predetermined breaking points due to the presence of channel-like structures underlying ridges in the testa [57]. In pea seeds the presence of xylogalacturonan in the inner walls of the testa was described, which coincides detached cells or with the junction sites between cells that are destined to detach, as described in other plant tissues [37]. In *Lepidium* seeds, HG composition shifts at the onset of testa rupture: while the seeds imbibe, non-esterified HG is ubiquitously distributed in all tissues, but by the time of testa rupture, this non-esterified HG is detected mostly on the endosperm and testa, meanwhile esterified HG is detected in the endosperm [38].

The seed testa is composed of several layers of nonliving cells, and thus the regulation and enzymes that facilitate cell separation in the testa must come from the living tissues underneath. At the onset of testa rupture (~25 HAI), it is possible to identify 90 cell wall-related genes from the 501 upregulated genes (~18%) in the micropylar endosperm and 58 from 282 genes (~20%) upregulated in the radicle. Also, about 8 (~8%) and 5 (~4%) genes were downregulated in both tissues, respectively [17]. In *Lepidium* seeds, tissue-specific transcript abundance patterns between the micropylar endosperm and the radicle accompany testa rupture, further supporting the view of this process as a decisive step in germination and in the regulation of cell wall-related genes [38]. In **Table 1**, these transcripts and its predicted biological/biochemical function are enlisted.

			Endosperm (HAI)				**RA**				**Endosperm (HAI)**				**RA**
Function	Gene ID	TS	0–12	16	25	31	25	Function	Gene ID	TS	0–12	16	25	31	25
EXPA10	AT1G26770	No	6–12		24		TR*	β-Gal	AT1G45130	EM	Not specified				
EXPA1	AT1G69530	No	3				TR*	β-Gal	AT5G08380	No	Not specified				
EXPA15	AT2G03090	No	12		24		TR*	β-Gal	AT1G77410	EN			24	31c	
EXPA6	AT2G28950	EM					TR	β-Gal	AT2G28470	No		16c	24		24

		Endosperm (HAI)					RA			Endosperm (HAI)					RA
EXPA4	AT2G39700	No		16				β-Gal	AT4G26140	No				31	
EXPA2	At5g05290	No	3		24			β-Glu	AT1G61820	EN	Not specified				
EXPA8	At2g40610	No	3–12		24			β-Glu	AT1G70710	No		16	TR*		
EXPA3	At2g37640	No	6–12		24			β-Glu	AT4G16260	EN	Not specified				
EXPA9	AT5G02260	No	12		24			β-Glu	AT1G26560	EM	Not specified				
EXPA20	AT4G38210	No	6–12		24			β-Glu	AT3G62750	No	3c				
EXPB1	At2g20750	No	12					β-Glu	AT2G44450	No			TR	31	
EXLA3	AT3G45960	No			TR			β-Glu	AT4G34480	No	12*				
EXLA1	AT3G45970	No		16	TR		TR	MAN5	AT4G28320	No	Not specified				
EXLA2	AT4G38400	No	6–12*		TR		TR	MAN6	AT5G01930	EN	Not specified				
EXT3	AT1G21310	EM	Not specified					MAN7	AT5G66460	No	6		24		
EXT10	AT5G06640	EM	Not specified					GH	AT5G49360	EN			TR		
EXTL	AT3G54590	EM	Not specified					GH	AT5G57560	No		16	TR		
EXTL	AT4G38770	EM	Not specified					GH	AT5G08370	No		16c	TR	31c	
EXTL	AT2G27380	RA	Not specified					GH	AT3G55430	No			TR	31	24*
XTH5	AT5G13870	No	6		24		24	GH	AT3G07320	No			24	31	24
XTH33	AT1G10550	No	12	16	TR		TR	GT	AT3G10320	EN				31c	
XTH	AT1G11545	No		16			TR	GH	AT3G13790	No		16		31	
XTH	AT1G32170	No			TR			GH	AT5G64570	No			24		TR
XTH17	AT1G65310	No			TR			GH-DUF3357	AT1G12240	No				31	TR
XTH	AT2G06850	No		16	TR	31	TR	GH	AT3G47010	EN			24		
XTH	AT2G36870	No				31		KOR2	AT1G65610	No				31	
XTH	AT3G23730	No			TR		TR	CESA5	AT5G09870	No			TR	31	24
XTH11	AT3G48580	No		16	24*	31		CSLC	AT4G07960	No			TR		TR
XTH	AT4G03210	No			TR		TR	AGP	AT3G11700	No			TR		TR
XTH	AT4G14130	No			TR		TR	AGP	AT5G44130	EN			TR		
XTH24	AT4G30270	EN		16	TR		TR	AGP	AT1G28290	No		16		31	
XTH18	AT4G30280	No		16	TR		TR	PRT	AT3G54400	No		16			24
XTH	AT4G30290	No		16		31		PRT	AT3G61820	No			TR*	31	24
XTH	AT4G37800	EM	Not specified					PRT	AT4G16563	No			TR		TR
XTH25	AT5G57550	No	6*		TR		TR	PL	AT3G24670	EM					24
XTR8	AT3G44990	EN		16c	24			PL	AT3G27400	No			TR		24
XTR6	AT4G25810	No			TR		TR	PL	AT4G13710	EN	Not specified				
PL	AT4G24780	No			TR*			PMEI	AT5G20740	No			TR		TR

			Endosperm (HAI)			RA				Endosperm (HAI)			RA
PL	AT5G48900	EM	Not specified				PMEI	AT5G46940	EN		TR		
PX	AT1G14540	EN		TR			PMEI	AT5G62340	No	16			TR
PX	AT1G14550	EN	Not specified				PMEI	AT5G64620	No			31c	
PX	AT1G30870	EM	Not specified				PG	AT3G59850	No	16	TR*	31	
PX	AT2G18980	No		TR	31		PG	AT3G61490	EM	Not specified			24
PX	AT2G43480	RA	Not specified				PG	AT4G23820	No 12		TR		24
PX	AT3G01190	EM	Not specified				Kinase	AT1G33590	No		TR	31	TR
PX	AT3G21770	EN			31c		Kinase	AT2G23770	No		TR		
PX	AT3G28200	No		TR	31		Kinase-DUF26	AT3G22060	MI	16		31	
PX	AT4G08770	EM	Not specified				Kinase	AT1G51940	No		TR		TR
PX	AT4G31760	EM	Not specified				LRR-p	AT4G26690	No		TR		TR
PX	AT5G05340	No			31		LRR-p	AT5G16590	No	16	TR	31	TR
PX	AT5G39580	No 3		TR			LRR-p	AT2G34930	No	16	TR	31	TR
PX	AT5G40150	EM	Not specified				DUF642	AT1G80240	EM	Not specified			
PX	AT5G64100	No	16	TR	31		DUF642	AT2G34510	EM	Not specified			
PX	AT5G64120	EN		TR	31		DUF642	AT2G41800	RA	Not specified			
PME	AT3G14310	No 6–12		24		24	DUF642	AT3G08030	No 3			31	
PME	AT1G04680	No			31	TR	DUF642	AT4G32460	No	16		31	
PME	AT1G57590	EM	Not specified			24	DUF642	AT5G11420	No 3			31	
PME	AT3G09410	EM	Not specified				DUF642	AT5G14150	EM	Not specified			
PME	AT3G10720	No	16	TR	31	TR	OX	AT1G62380	No	16c			
PME	AT3G62060	EM	Not specified			24	OX	AT1G76160	No		TR		
PME	AT4G19420	EN	Not specified				OX	AT2G46740	EN	16	TR	31	
PME	AT5G26670	RA	Not specified				OX	AT4G22010	No		TR		
PME	AT5G45280	No 3			31		OX	AT4G38420	No		TR		TR
PME	AT5G62330	EM	Not specified			24	OX	AT5G21105	No	16	TR	31	
PME2	AT1G02810	No 12	16		31		OX	AT5G44380	No		TR	31	
PME2	AT1G11580	No	16			24	PTRI	AT1G17860	No		TR	31	TR
PME2	AT2G26440	No	16		31	24	PTRI	AT2G38870	No			31	TR
PME2	AT3G47400	No			31	24	PTRI	AT4G22470	MI		TR	31	
PME2	AT3G49220	No		TR		TR	GT	AT1G64390	No	16	TR*	31	
PME2	AT4G02330	No	16	TR	31	TR	GT	AT2G02990	EN		TR		
PME2	AT4G33220	EM	Not specified				GT	AT2G14610	EN	Not specified			
PME2	AT5G64640	EM 12		24			GT	AT1G05170	EM				TR*

			Endosperm (HAI)	RA				Endosperm (HAI)	RA
PMEI	AT1G62770	RA	Not specified		GT	AT1G08280	No	31	
PMEI	AT2G47670	EM	3-12* 24*		PGIP	AT5G06860	No		TR
PMEI	AT4G00080	No	31	TR	PG	AT2G43860	MI	24 31	
PMEI	AT4G12390	EM	Not specified		PG	AT3G06770	No	TR	24

The endosperm expression profiles are subdivided by hours after imbibition (HAI) and testa rupture (TR, ~25HAI), whereas the radicle (Rad) expression only shows the moment of TR. *Abbreviations*: T-S, tissue-specific expression; EM, embryo; EN, endosperm; RA, radicle; MI, micropylar endosperm; EXPA, expansin; EXT, extensin; EXTL, extension-like protein; XTH, xyloglucan-transglycosylhydrolase; GH, glycosyl-hydrolase; β-Glu, β-glucosidase; β-Gal, β-galactosidase; AGP, arabinogalactan protein; CESA, cellulose synthase; PRT, protease; PL, pectate lyase; PX, peroxidase; CSLC, cellulose synthase-like; LRR-p, leucine-rich repeat protein; DUF, domain of unknown function; PME, pectin methylesterase; PME2, PME with an inhibitory domain; PMEI, PME inhibitor; PTRI, protease inhibitor; GT, transglycanase; PG, polygalacturonase; PGIP, PG inhibitor. The * means downregulation for that particular gene and time. The letter "c" beside a number means the expression was upregulated in the chalazal endosperm.

Table 1. Expression of some upregulated cell wall-modifying genes in *Arabidopsis* tissue-specific microarrays described by [17, 22] and expression profiles at http://bar.utoronto.ca/.

In *Arabidopsis* and *Lepidium* seeds, total PME activity increases gradually with imbibition time and peaks at the onset of testa rupture. ABA treatment does not affect testa rupture but endosperm rupture is delayed and PME activity fails to decrease following testa rupture [16, 38]. Two DUF642 genes, *BIIDXI* (*BDX, At4g32460*) and *At5g11420*, are expressed in the embryo and micropylar endosperm, respectively, during germination. In overexpression lines, testa and endosperm rupture of matrix-primed seeds occurred earlier compared to wild type. The germination performance of overexpression seeds was accompanied by an increase of total PME activity, compared to the wild type [50].

In non-endospermic seeds, testa rupture marks the end of germination. In this type of seeds, the testa rupture is accompanied by radicle elongation whose continued pressure in the inner face of the testa promotes cell separation [53].

3.4.3. *Endosperm weakening for radicle protrusion*

The endosperm functions as a barrier to control radicle protrusion as it can impose primary dormancy in many species like *Arabidopsis, Lepidium,* and yellow cedar, among others [2]. Endosperm structure of mature seeds varies greatly within species, where it can comprise one layer of cells as in *Arabidopsis, Lepidium,* and cucumber or to several layers as in tomato or tobacco [58, 59]. Structural studies in hard-seeded species like fenugreek (*Trigonella foenum-graecum*) and coffee (*Coffea arabica*) have demonstrated that near the micropyle a zone of thin-walled cells that can be a low resistance area for radicle protrusion exists. Endosperm cell wall composition varies considerably among species: in the closely related species, *Arabidopsis* and *Lepidium* are rich in cellulose, non-esterified HG, arabinans, and XG. However, these polysaccharides are not uniformly distributed in the endosperm: in *Lepidium*, an epitope for arabinans (LM13) was localized in the inner and outer walls of the cells, but absent from the traverse walls. The endosperm of tomato seed contains mannans that have been shown to contribute

http://dx.doi.org/10.5772/64085

to the control of radicle protrusion and general endosperm hardness rather than a storage function [39].

The hydrolytic activity of β-glucanases or endo-β-mannanases can contribute to endosperm cell wall weakening in Brassicaceae and Solanaceae species, which have cell walls rich in mannans [59]. In *Arabidopsis* seeds the activity of *AtMAN5, AtMAN6,* and *AtMAN7* and regulation of their activity by the basic leucine-zipper 44 transcription factor, *AtbZIP44,* whose knockout mutants have delayed germination have been described [60]. In hard-seeded species, there was a negligible activity in the radicle and micropylar endosperm of endo-β-mannanases that could not be associated with endosperm weakening to allow radicle protrusion [61]. However it is still unexplored the activity of other CWMPs that could contribute to wall rearrangements during germination in these species. The upregulation of wall-related genes in the endosperm has been reported; this includes α-expansins (*AtEXPA2, AtEXPA8,* and *AtEXPA9*), β-expansin (*AtEXPB1*), expansin-like protein (*AtEXPL1*), cellulose synthase-like proteins (*CSLA2* and *CSLC4*), xyloglucan endotransglycosylases (*AtXTH11, AtXTH17, AtXTH18, AtXTH33, AtXTH31, AtXTH23,* and *AtXTH24*), and mannanase (*AtMAN7*) [62].

Several reports indicate that endosperm weakening and rupture are inhibited or delayed by ABA, in a dose-dependent manner, in some species of the Brassicaceae family [8, 16] and tobacco [63]. Microarray analysis of *Arabidopsis* seeds treated with exogenous ABA at the onset of radicle protrusion has shown a downregulation of several wall-related transcripts in the endosperm, including PMEs, AGPs, and PLs [7]. In tobacco, ABA delays the accumulation and activity of β-1,3-glucanase in the micropyle before radicle protrusion [63]. PMEs contribute to seed germination in several species by modulating the degree of methylesterification of pectins in the endosperm [16, 38, 64, 65]. In yellow cedar a loss of the internal structure of the megagametophyte surrounding the radicle during germination was described. The resulting decrease in the mechanical strength of the megagametophyte would allow radicle protrusion. There is a positive correlation between dormancy alleviation and PME activity, as well as with germination performance and PME activity in both the megagametophyte and the embryo. PME activity has also been demonstrated to positively correlate with germination performance in *Arabidopsis* [64]. The *LeXET4* gene transcripts, which are restricted to the endosperm cap, are detected in tomato seeds within 12 h of imbibition and reach a maximum at 24 h [66]; they decline after radicle emergence despite a continued degradation of the lateral endosperm cell walls. PG *At2g43860* is expressed within the endosperm cells of the seed adjacent to the site of the emerging radicle [67]. In germinating tomato seeds, several reports indicate the expression of expansins [51], PGs (*LeXPG1*) [68], and XTHs (*LeXET4*), which accumulate in the endosperm region adjacent to the expanding radicle (~40HAI). Transcripts are detected within 12HAI and generally peak by 24HAI, consistent with the endosperm weakening.

3.4.4. Embryonic axis elongation

Cell elongation, rather than cellular division, is the main process that drives embryo growth [14]. Cell division occurs after radicle protrusion and contributes to the rapid growth of the embryonic axis by generating new elongating cells [2]. Cell elongation that drives radicle protrusion occurs at the transition zone, which comprises the cells between the last proximal root hair cell in the radicle and the lower basal cells of the hypocotyl [69]. In *Arabidopsis* Col seeds that have been previously stratified, the radicle protrusion can initiate as early as 32 HAI. By this time, and immediately prior to the radicle emergence through the endosperm, the cells in the transition zone had incremented their size by 44% while the cells in the radicle 10% and in the hypocotyl 30%. By 40HAI, the radicle has already protruded and the elongated cells in the seedling have increased their size by 15% in the radicle, 52% in the hypocotyl, and 108% in the transition zone. Elongation is often accompanied by an increase in DNA content without subsequent mitosis (endoreduplication) [69].

In the micropylar endosperm of tomato seeds, an important mobilization of protein bodies occurs, but it seems that there is no cell degradation as the radicle protrudes. Instead, a process similar to cell separation to allow radical protrusion was suggested [70]. A similar process was observed in celery seeds, where the radicle tip also seems to penetrate the micropylar endosperm by separating the endosperm cells, but, since the embryo needs to grow before germination is completed, cell degradation for storage mobilization occurs in the endosperm adjacent to the embryo [27]. The expression of *LeEXP8* and *LePG1* in the embryo elongation zone of tomato seeds has been reported at the onset of radicle protrusion [51, 68].

3.4.5. Cell wall participation in the mobilization of stored reserves

Major reserve mobilization occurs once germination has concluded, and these reserves are utilized to feed the growing seedling rather than to fuel radicle protrusion. However, in cereal grains, the preparation for starch and oligosaccharide mobilization occurs within the first hours of germination [15]. In cereals, the endosperm is a nonliving storage tissue, and the endosperm cell walls protect its contents from enzymatic attack. Accordingly, the degradation of cell walls is a limiting step in storage reserve mobilization that is induced by the GA produced by the embryo (at the scutellum) and secreted to the aleurone layer [2].

In most endospermic seeds this tissue is still living. Mannans in the endosperm cell walls of date palm (*Phoenix dactylifera*) and coffee are mobilized to support embryo development. It has been proposed that the mobilization of storage xyloglucans can be coupled to the growth rate of the seedling by transglycosylation. In legumes, endosperm galactomannans seems to function as reserves; they can constitute up to 30% of total seed dry weight. In fenugreek and *Schizolobium parahyba*, the cell walls of the endosperm are thickened with galactomannan and in some cases the cytoplasm is nonexistent. In *Tamarindus indica* and *Hymenaea courbaril*, reserve xyloglucans are stored between two primary walls and are degraded without hydrolyzing both walls [39]. During germination of celery seeds, the surrounding endosperm degrades leaving a small amount of un-degraded polymers of the cell wall, except for the micropylar endosperm, in which only some protein bodies are mobilized and the rest of the cells persist until the radicle pushes through; once radicle protrusion has started, these micropylar cells are degraded [27].

4. Concluding remarks

In-depth temporal screening of cell wall-related transcripts and proteins has provided an important overview of the possible actors involved in the five stages during germination where wall modification is involved, as described above. In *rice*, a valuable integrative effort using -omics approaches has been done to understand seed germination [23]. This analysis suggested that the changes in transcript levels during early germination (3–12 HAI) drive the subsequent changes in the metabolome (12–24 HAI) of germinating seeds, supporting that most of the changes observed at the transcriptional level are related to the cellular processes involved in germination. Other authors have associated transcript abundance with specific seed compartments and some enzymatic activity assays, demonstrating the relevance of understanding tissue-specific expression profiles [17, 38]. Much of the information available related to CWMPs still needs to be validated through enzymatic activity or in vivo interaction assays. Only about 121 (~12%) of the total cell wall-related genes are experimentally validated [48]. Also, many cell wall-related proteins belong to families of unknown function. The -omics approach can be useful to propose hypothesis of wall-modification complexes, whose activity could be regulated at several levels, and coordinated by unknown function proteins that could act as scaffolding proteins and direct this complex activity to specific polysaccharides. Cosgrove [44] proposed that CWMPs could be functionally classified into primary or secondary modifiers, but this idea has not being reflected in other studies. Following Cosgrove, the analysis of cell wall modification considering an alternating activity of primary or secondary modifiers could facilitate the understanding of the dynamics of cell wall modification during seed germination. For instance, expansins could be primary modifiers as they affect cell wall loosening and extensibility, but they do not remove or transfer polysaccharides into the wall during imbibition; other primary modifiers could be PMEs, as their activity precedes PGs and promotes cell expansion or cell separation, or the resulting exposed GAL residues can be cross-linked with Ca^{2+} and promote wall stiffening. Secondary modifiers would include GHs and GTs that would act on exposed residues either promoting cell expansion, separation, or stiffening. In assays to study mucilage properties, sequential treatment with different hydrolases allows solubilization of other components, which are masked to the activity of other enzymes [55, 71]. The alternate perspective of primary and secondary modifiers could help in identifying potential interactions in silico and tested in vivo.

Spatial transcriptomic analyses that include the different seed compartments and the analysis of cell wall composition changes using specific antibodies for in situ localization of the different polysaccharide epitopes in seed tissues provide valuable information. Although *Arabidopsis* is the best-known plant model, several authors demonstrate that comparative analysis allows higher resolution of tissue-specific cell wall microdomains that are not achievable in *Arabidopsis* [8, 10, 60]. As an example, Lee et al. [59] describe the presence of LM13 epitopes in the inner and outer cell walls, but absent in the transverse cell walls of the endosperm in *Lepidium* seeds; in tobacco, which has a thicker endosperm than *Arabidopsis* or *Lepidium*, XGs were abundant in the embryo, and at the micropyle (rich in heteromannans), these polysaccharides were only present in the middle lamella and intercellular regions. Thus, the analysis of cell wall-modification processes would benefit from the multispecies comparison of in situ localization of polysaccharide epitopes in seed tissues. The characterization of wall microdomains could be combined with the valuable information generated by -omics technologies, to propose new hypothesis of regulation and coordinated activity of CWMPs. Ultimately, the activity of these CWMPs must be confirmed by in situ localization, in vivo protein interactions, and enzymatic activity. By combining the resources available for model species with the selection of other plant systems with bigger-easy-to-handle seeds, it could be possible to achieve a comprehensive view of seed-compartment functions and regulation during germination. The endosperm role during germination is fundamental in endospermic seeds; however, in non-endospermic seeds, this role must befall on either the embryo or the testa. Since the testa is a nonliving tissue, the radicle most certainly assumes part of this regulatory role, but a comparative analysis is needed to ascertain this supposition and to determine if some of the endosperm functions are developed by the testa while still in the maturation program. The occurrence of endospermic and non-endospermic seeds within the same taxa is relatively common in legumes such as *soybean*, which could offer a model for analyzing transcriptomic differences within embryo compartments comparable to the differences described between the endosperm and the radicle.

ROS participation in germination is supported by several reports and transcriptomic profiles of germinating seeds [9, 14, 17]. However, the actual role of ROS and ROS-related enzymes in promoting cell wall loosening needs to be further analyzed, since the physiological concentrations of ROS during germination do not seem to be sufficient to induce wall extension, and attempts of increasing ROS concentration lead to wall breakage [5]. Müller et al. [72] describe abnormal rupture of the micropylar endosperm of *Lepidium* seeds treated with H_2O_2, while the treatment with myrigalone A [73], which inhibits the hormone-mediated accumulation of ROS during germination, also induces abnormal endosperm breakage. These observations further support the notion of ROS as a signaling agent that induces downstream activation of CWMPs than inducing wall loosening on its own.

Acknowledgements

This work was supported by Programa de Apoyo a Proyectos de Investigación e Innovación Tecnológica (PAPIIT) IN207915 and SEP-CONACyT grant 155074. Ximena Gómez-Maqueo acknowledges the scholarship by the Consejo Nacional de Ciencia y Tecnología (CONACyT).

Author details

Ximena Gómez-Maqueo and Alicia Gamboa-deBuen*

*Address all correspondence to: agamboa@ecologia.unam.mx

Ecological Physiology Lab., Ecology Institute, National Autonomous University of Mexico, UNAM, Mexico City, Mexico

References

[1] Baskin CC, Baskin JM. Seeds: ecology, biogeography, and evolution of dormancy and germination. San Diego, Elsevier; 1998. 666 p. ISBN-13: 978-0120802609

[2] Bewley JD, Bradford KJ, Hilhorst HWM, Nonogaki H. Seeds, physiology of development, germination and dormancy. 3rd ed. New York, Springer; 2013. 392p. DOI 10.1007/978-1-4614-4693-4.

[3] Fenner M, Thompson K. The ecology of seeds. Cambridge, Cambridge University Press; 2005. 250p.

[4] Taiz L, Zeiger E. Plant physiology. 5th Ed. Sunderland: Sinauer Assoc; 2010. 782p. ISBN-13: 978-0878938667

[5] Cosgrove DJ. Growth of the plant cell wall. Nature Reviews Molecular Cell Biology. 2005;6:850-861. DOI: 10.1038/nrm1746

[6] Donohue K, Dorn L, Griffith C, Kim E, Aguilera A, Polisetty CR, Schmitt J. The evolutionary ecology of seed germination of *Arabidopsis thaliana*: variable natural selection on germination timing. Evolution. 2005;59:758-770. DOI: 10.1111/j.0014-3820.2005.tb01751.x

[7] Penfield S, Li Y, Gilday AD, Graham S, Graham IA. Arabidopsis ABA INSENSITIVE4 regulates lipid mobilization in the embryo and reveals repression of seed germination by the endosperm. The Plant Cell. 2006;18:1887-1899. DOI: http://dx.doi.org/10.1105/tpc.106.041277

[8] Müller K, Tintelnot S, Leubner-Metzger G. Endosperm-limited Brassicaceae seed germination: abscisic acid inhibits embryo-induced endosperm weakening of *Lepidium sativum* (cress) and endosperm rupture of cress and *Arabidopsis thaliana*. Plant and Cell Physiology. 2006;47:864-877. DOI: 10.1093/pcp/pcj059

[9] Linkies A, Leubner-Metzger G. Beyond gibberellins and abscisic acid: how ethylene and jasmonates control seed germination. Plant Cell Reports. 2012;31:253-270. DOI: 10.1007/s00299-011-1180-1

[10] Linkies A, Müller K, Morris K, Turečková V, Wenk M, Cadman CS, Corbineau F, Strnad M, Lynn JR, Finch-Savage WE, Leubner-Metzger G. Ethylene interacts with abscisic acid to regulate endosperm rupture during germination: a comparative approach using *Lepidium sativum* and *Arabidopsis thaliana*. The Plant Cell. 2009;21:3803-3822. DOI: http://dx.doi.org/10.1105/tpc.109.070201

[11] Kanai M, Nishimura M, Hayashi M. A peroxisomal ABC transporter promotes seed germination by inducing pectin degradation under the control of ABI5. The Plant Journal. 2010;62:936-947. DOI: 10.1111/j.1365-313X.2010.04205.x

[12] Bewley JD. Seed germination and dormancy. The Plant Cell. 1997;9(7):1055. Plant Cell. 1997;9:1055–1066. DOI: 10.1105/tpc.9.7.1055

[13] Nonogaki H, Bassel GW, Bewley JD. Germination—still a mystery. Plant Science. 2010;179:574-581. DOI: 10.1016/j.plantsci.2010.02.010

[14] Weitbrecht K, Müller K, Leubner-Metzger G. First off the mark: early seed germination. Journal of Experimental Botany. 2011;62:3289-3309. DOI: 10.1093/jxb/err030

[15] Rosental L, Nonogaki H, Fait A. Activation and regulation of primary metabolism during seed germination. Seed Science Research. 2014;24:1-5. DOI: http://dx.doi.org/10.1017/S0960258513000391

[16] Müller K, Levesque-Tremblay G, Bartels S, Weitbrecht K, Wormit A, Usadel B, Haughn G, Kermode AR. Demethylesterification of cell wall pectins in Arabidopsis plays a role in seed germination. Plant Physiology. 2013;161:305-316. DOI: http://dx.doi.org/10.1104/pp.112.205724

[17] Dekkers BJ, Pearce S, van Bolderen-Veldkamp RP, Marshall A, Widera P, Gilbert J, Drost HG, Bassel GW, Müller K, King JR, Wood AT. Transcriptional dynamics of two seed compartments with opposing roles in Arabidopsis seed germination. Plant Physiology. 2013;163:205-215. DOI: http://dx.doi.org/10.1104/pp.113.223511

[18] Yan D, Duermeyer L, Leoveanu C, Nambara E. The functions of the endosperm during seed germination. Plant Cell and Physiology. 2014;55:1521-1533. DOI: 10.1093/pcp/pcu089

[19] Narsai R, Law SR, Carrie C, Xu L, Whelan J. In-depth temporal transcriptome profiling reveals a crucial developmental switch with roles for RNA processing and organelle metabolism that are essential for germination in Arabidopsis. Plant Physiology. 2011;157:1342-1362. DOI: http://dx.doi.org/10.1104/pp.111.183129

[20] Jia H, Suzuki M, McCarty DR. Regulation of the seed to seedling developmental phase transition by the LAFL and VAL transcription factor networks. Wiley Interdisciplinary Reviews: Developmental Biology. 2014;3:135-145. DOI: 10.1002/wdev.126

[21] Kermode AR, Finch-Savage BE. Desiccation sensitivity in orthodox and recalcitrant seeds in relation to development. In: Black M, Pritchard HW, editors. Desiccation and survival in plants: drying without dying. New York: Cabi; 2002; p:149-184

[22] Nakabayashi K, Okamoto M, Koshiba T, Kamiya Y, Nambara E. Genome-wide profiling of stored mRNA in *Arabidopsis thaliana* seed germination: epigenetic and genetic regulation of transcription in seed. The Plant Journal. 2005;41:697-709. DOI: 10.1111/j.1365-313X.2005.02337.x

[23] Howell KA, Narsai R, Carroll A, Ivanova A, Lohse M, Usadel B, Millar AH, Whelan J. Mapping metabolic and transcript temporal switches during germination in rice highlights specific transcription factors and the role of RNA instability in the germi-

nation process. Plant Physiology. 2009;149:961-980. DOI: http://dx.doi.org/10.1104/pp.108.129874

[24] Ogawa M, Hanada A, Yamauchi Y, Kuwahara A, Kamiya Y, Yamaguchi S. Gibberellin biosynthesis and response during Arabidopsis seed germination. The Plant Cell. 2003;15:1591-1604. DOI: http://dx.doi.org/10.1105/tpc.011650

[25] Silverstone AL, Chang CW, Krol E, Sun TP. Developmental regulation of the gibberellin biosynthetic gene GA1 in *Arabidopsis thaliana*. The Plant Journal. 1997;12:9-19. DOI: 10.1046/j.1365-313X.1997.12010009.x

[26] Groot SP, Karssen CM. Gibberellins regulate seed germination in tomato by endosperm weakening: a study with gibberellin-deficient mutants. Planta. 1987;171:525-531.

[27] Jacobsen JV, Pressman E. A structural study of germination in celery (*Apium graveolens* L.) seed with emphasis on endosperm breakdown. Planta. 1979;144:241-248.

[28] Pelloux J, Rusterucci C, Mellerowicz EJ. New insights into pectin methylesterase structure and function. Trends in Plant Science. 2007;12:267-277. DOI: 10.1016/j.tplants.2007.04.001

[29] Caffall KH, Mohnen D. The structure, function, and biosynthesis of plant cell wall pectic polysaccharides. Carbohydrate Research. 2009;344:1879-1900. DOI: 10.1016/j.carres.2009.05.021

[30] Franková L, Fry SC. Biochemistry and physiological roles of enzymes that 'cut and paste' plant cell-wall polysaccharides. Journal of Experimental Botany. 2013;64:3519–3550. DOI: 10.1093/jxb/ert201

[31] Levesque-Tremblay G Pelloux J, Braybrook SA. Tuning of pectin methylesterification: consequences for cell wall biomechanics and development. Planta. 2015; 242(4):791–811. DOI 10.1007/s00425-015-2358-5

[32] Carpita NC. Structure and biogenesis of the cell walls of grasses. Annual Review of Plant Biology. 1996;47:445-476. DOI: 10.1146/annurev.arplant.47.1.445

[33] Yokoyama R, Nishitani K. Genomic basis for cell-wall diversity in plants. A comparative approach to gene families in rice and Arabidopsis. Plant and Cell Physiology. 2004;45:1111-1121. DOI: 10.1093/pcp/pch151

[34] Reiter W. Biosynthesis and properties of the plant cell wall. Current Opinion in Plant Biology. 2002;5:536-542. DOI: 10.1016/S1369-5266(02)00306-0

[35] Bashline L, Li S, Gu Y. The trafficking of the cellulose synthase complex in higher plants. Annals of Botany. 2014;114:1059-1067. DOI: 10.1093/aob/mcu040

[36] Mohnen, D. Pectin structure and biosynthesis. Current Opinion in Plant biology. 2008;11:266-277. DOI: 10.1016/j.pbi.2008.03.006

[37] Willats WG, McCartney L, Steele-King CG, Marcus SE, Mort A, Huisman M, van Alebeek GJ, Schols HA, Voragen AG, Le Goff A, Bonnin E. A xylogalacturonan epitope

[61] Gong X, Bassel GW, Wang A, Greenwood JS, Bewley JD. The emergence of embryos from hard seeds is related to the structure of the cell walls of the micropylar endosperm, and not to endo-β-mannanase activity. Annals of Botany. 2005;96:1165-1173. DOI: 10.1093/aob/mci269

[62] Endo A, Tatematsu K, Hanada K, Duermeyer L, Okamoto M, Yonekura-Sakakibara K, Saito K, Toyoda T, Kawakami N, Kamiya Y, Seki M. Tissue-specific transcriptome analysis reveals cell wall metabolism, flavonol biosynthesis and defense responses are activated in the endosperm of germinating *Arabidopsis thaliana* seeds. Plant and Cell Physiology. 2012;53:16-27. DOI: 10.1093/pcp/pcr171

[63] Leubner-Metzger G, Frundt C, Vogeli-Lange R, Meins Jr F. Class I [β]-1, 3-Glucanases in the endosperm of tobacco during germination. Plant Physiology. 1995;109:751-759. DOI: http://dx.doi.org/10.1104/pp.109.3.751

[64] Ren C, Kermode AR. An increase in pectin methyl esterase activity accompanies dormancy breakage and germination of yellow cedar seeds. Plant Physiology. 2000;124:231-242. DOI: http://dx.doi.org/10.1104/pp.124.1.231

[65] Martínez-Andújar C, Pluskota WE, Bassel GW, Asahina M, Pupel P, Nguyen TT, Takeda-Kamiya N, Toubiana D, Bai B, Górecki RJ, Fait A. Mechanisms of hormonal regulation of endosperm cap-specific gene expression in tomato seeds. The Plant Journal. 2012;71:575-86. DOI: 10.1111/j.1365-313X.2012.05010.x

[66] Chen F, Nonogaki H, Bradford KJ. A gibberellin-regulated xyloglucan endotransglycosylase gene is expressed in the endosperm cap during tomato seed germination. Journal of Experimental Botany. 2000;53:215-223. DOI: 10.1093/jexbot/53.367.215

[67] González-Carranza ZH, Elliott KA, Roberts JA. Expression of polygalacturonases and evidence to support their role during cell separation processes in *Arabidopsis thaliana*. Journal of Experimental Botany. 2007;58:3719-3730. DOI: 10.1093/jxb/erm222

[68] Sitrit Y, Hadfield KA, Bennett AB, Bradford KJ, Downie AB. Expression of a polygalacturonase associated with tomato seed germination. Plant Physiology. 1999;121:419-428. DOI: http://dx.doi.org/10.1104/pp.121.2.419

[69] Sliwinska E, Bassel GW, Bewley JD. Germination of *Arabidopsis thaliana* seeds is not completed as a result of elongation of the radicle but of the adjacent transition zone and lower hypocotyl. Journal of Experimental Botany. 2009;60:3587-3594. DOI: 10.1093/jxb/erp203

[70] Karssen CM, Haigh A, Van der Toorn P, Weges R. Physiological mechanisms involved in seed priming. In. Taylorson, R.B. Recent advances in the development and germination of seeds. 1989 (pp. 269-280). New York, Springer US.

[71] North HM, Berger A, Saez-Aguayo S, Ralet MC. Understanding polysaccharide production and properties using seed coat mutants: future perspectives for the

exploitation of natural variants. Annals of Botany. 2014;114:1251-1263. DOI: 10.1093/aob/mcu011

[72] Müller K, Linkies A, Vreeburg RA, Fry SC, Krieger-Liszkay A, Leubner-Metzger G. In vivo cell wall loosening by hydroxyl radicals during cress seed germination and elongation growth. Plant Physiology. 2009;150:1855-65. DOI: http://dx.doi.org/10.1104/pp.109.139204

[73] Oracz K, Voegele A, Tarkowská D, Jacquemoud D, Turečková V, Urbanová T, Strnad M, Sliwinska E, Leubner-Metzger G. Myrigalone A inhibits *Lepidium sativum* seed germination by interference with gibberellin metabolism and apoplastic superoxide production required for embryo extension growth and endosperm rupture. Plant and Cell Physiology. 2012;53:81-95. DOI: 10.1093/pcp/pcr124

Chapter 8

Seed Germination Technologies for Helophyte Production Used in Wastewater Treatment

Trinidad Ruiz-Téllez, Francisco M. Vázquez-Pardo,
José Blanco-Salas and F. Javier Carbonell-Espín

Additional information is available at the end of the chapter

http://dx.doi.org/10.5772/64084

Abstract

Constructed green wetlands with horizontal surface for wastewater treatment are gaining acceptance. Many countries have published innovative experiences with this technology. A great variety of wastewaters from industries have been treated. Different plant species have been tested. Seed technology development provides interesting tools to produce these species in nurseries. It is a sustainable new business. But studies on seed germination of aquatic and lacustrine plants are very few. That is why we have made the following bibliographic review. We have summarised and analysed the state of the art of this innovative topic, concluding that seed technology for multiplication of helophytes needs further experimental work. But there is enough information to produce right now, tens of different species. Significant efforts have been done. Even though it is a challenge to produce from now on, experimental results are ready to be transferred to those who are trading with this type of plants. Helophytes have a promising future as sustainable elements of the upcoming sewage equipment. Improvements on the biotechnology of these species are a worthwhile researching line. To this aim, the following revision is an essential compilation with which to begin.

Keywords: germination protocol, green filter, helophyte, wastewater, production

1. Introduction

Constructed green wetlands with horizontal surface for wastewater treatment are getting social recognition. They usually have plants as an essential component of the design. Many countries have published innovative experiences with this technology: the USA, Canada, the UK,

Germany, France, the Netherlands, Switzerland, Norway, Poland, Slovenia, Lithuania, Italy, Spain, Portugal, Australia, Japan, China, India, Taiwan, South Africa, Turkey, Kenya, Uganda and México [1, 2]. A great variety of wastewaters from industries have been treated: chemical, petrochemical, textile, pulp and paper, tannery, abattoir, food processing, distillery and winery factories. Effluents from pig farms, fish farms, shrimp culture, cobalt recovery, mining or coke plants have also been managed this way. Run-offs from airports, highways, hospitals, agricultural activities or storm water have been tested as well. Finally, landfill leachate or polluted rivers have been experimentally decontaminated using constructed wetlands with vascular plants as the principal component [1, 3]. Uptake of rare earth elements (REEs) from obsolete equipment of modern technologies has also been reported [4].

Different species have been used to date, all over the world. They belong to different genera and families. Frequently they are helophytes, which means plants growing in marsh partly submerged in water, so that they regrow from buds below the water surface. The helophytic plants used to build wastewater green filters in constructed wetlands have in common with these significant characteristics: (a) a rich below-ground organ, root or rhizome, as to provide substrate for attached bacteria and oxygenation (as much as possible) of areas adjacent to the radicular apparatus; (b) a high-tolerance nutrient and organic loadings; and (c) the production of a high amount of above-ground biomass for winter insulation in cold and temperate regions, as well as nutrient removal via harvesting [1].

From a botanical point of view, a great list of species could be used in each country, because plant biodiversity is high. Ad hoc species for each biogeographical region could be found, attending to the above-mentioned characteristics. However, in practical terms, a limited number of species have been tested by the industry. Vymazal [1] summarised the most important, and we have taken its publication as a framework. Later references [4–12] have also been consulted, and **Table 1** shows the helophytes that we have considered most relevant.

For the production and multiplication of plants in nurseries to the global market, it is necessary to standardise the operations aimed at getting living plants that can be installed in the constructed sewages. Plant reproduction can be made by vegetative way (cuttings) or by seeds (sexual way). Nowadays international trade is essential, so those systems better adapted to international transport and management are more competitive. For this reason seed multiplication systems are very interesting. Seeds are more resistant and better prepared than cuttings, to adverse environmental conditions (light, humidity, temperature), that can occur during international transport. Thus, seed technology development provides interesting tools to produce in nurseries good quality plants. It is a sustainable new business.

Various authors have highlighted that studies on seed germination of aquatic and lacustrine plants are very few [13]. This topic is a much less studied subject that other aspects of seed biology, physiology or ecology. However it has a great importance from an applied point of view. That is why we have made the following bibliographic review. It will let us summarise and analyse the state of the art of this interesting topic.

2. Seed germination conditions for the main helophyte groups for wastewater treatment

A prospective analysis of the publications issued in specialised databases allows us to know the main techniques proposed for seed germination, fitting us to the scope of (**Table 1**) helophyte species. We set out below the most relevant conditions for a successful germination of the genera and species internationally used, excluding palms and aquatic ornamentals.

Species	Family
Acorus calamus L.	Acoraceae
Arundo donax L.	Poaceae
Asclepias incarnata L.	Apocynaceae
Baumea articulata Gaudich.	Cyperaceae
Brachiaria mutica (Forssk.) Stapf	Poaceae
Canna glauca L.	Cannaceae
Canna indica L.	Cannaceae
Canna x generalis L. H. Bailey	Cannaceae
Carex acutiformis Ehrh.	Cyperaceae
Carex gracilis Curtis.	Cyperaceae
Carex lacustris Willd.	Cyperaceae
Coix lacryma-jobi L.	Poaceae
Colocasia esculenta (L.) Schott	Araceae
Cyperus esculentus L.	Cyperaceae
Cyperus alterniflorus Willd. ex Kunth	Cyperaceae
Cyperus articulatus L.	Cyperaceae
Cyperus dubius Rottb.	Cyperaceae
Cyperus flabelliformis L.	Cyperaceae
Cyperus grandis C. B. Clarke	Cyperaceae
Cyperus immensus C. Presl	Cyperaceae
Cyperus involucratus Rottb.	Cyperaceae
Cyperus isocladus L.	Cyperaceae
Cyperus malaccensis Lam.	Cyperaceae
Cyperus esculentus L.	Cyperaceae
Echinochloa polystachya (Kunth) A. S. Hitchc.	Poaceae
Eichhornia crassipes (Mart.) Solms-Laub.	Pontederiaceae
Eleocharis sphacelata R. Br.	Cyperaceae

Species	Family
Epilobium hirsutum L.	Onagraceae
Festuca arundinacea Schreb.	Poaceae
Filipendula ulmaria (L.) Maxim.	Rosaceae
Glyceria maxima (Hartm.) Holmb.	Poaceae
Gynerium sagittatum (Aubl.) P. Beauv	Poaceae
Heliconia psittacorum L. f.	Heliconiaceae
Heliconia rostrata Ruiz and Pav.	Heliconiaceae
Hemerocallis fulva L.	Xanthorrhoeaceae
Hibiscus moscheutos L.	Malvaceae
Hymenocallis littoralis (Jacq.) Salisbury	Amaryllidaceae
Iris pseudacorus L.	Iridaceae
Iris tectorum Maxim.	Iridaceae
Iris versicolor L.	Iridaceae
Juncus effusus L.	Juncaceae
Juncus inflexus L.	Juncaceae
Juncus subsecundus N. A. Wakef.	Juncaceae
Kyllinga erecta Schumach.	Cyperaceae
Lepironia articulata (Retz.) Domin	Cyperaceae
Liatris pycnostachya Michx.	Asteraceae
Lobelia cardinalis L.	Campanulaceae
Lythrum salicaria L.	Lythraceae
Mentha spicata L.	Lamiaceae
Monochoria vaginalis (Burm. f.) C. Presl ex Kunth	Pontederiaceae
Panicum maximum Jacq.	Poaceae
Panicum repens L.	Poaceae
Paspalum distichum L.	Poaceae
Pennisetum purpureum Schumach.	Poaceae
Phalaris arundinacea L.	Poaceae
Phragmites australis (Cav.) Trin. ex Steud.	Poaceae
Phragmites karka (Retz.) Trin. ex Steud.	Poaceae
Phragmites mauritianus Kunth	Poaceae
Phylidrum lanuginosum Banks and Sol. ex Gaertn.	Philydraceae
Pontederia cordata L.	Pontederiaceae
Rudbeckia hirta L.	Asteraceae

http://dx.doi.org/10.5772/64084

Species	Family
Sagittaria latifolia Willd.	Alismataceae
Scirpus acutus (Muhl. ex J. M. Bigelow) Á. Löve and D. Löve	Cyperaceae
Scirpus americanus Pers.	Cyperaceae
Scirpus californicus (C. A. Mey.) Steud.	Cyperaceae
Scirpus cyperinus (L.) Kunth	Cyperaceae
Scirpus fluviatilis (Torr.) A. Gray	Cyperaceae
Scirpus grossus L.	Cyperaceae
Scirpus lacustris (L.) Palla	Cyperaceae
Scirpus maritimus L.	Cyperaceae
Scirpus pungens Vahl	Cyperaceae
Scirpus sylvaticus L.	Cyperaceae
Scirpus tabernaemontani (C. C. Gmel.) Missbach and E. H. L. Krause	Cyperaceae
Scirpus validus Vahl	Cyperaceae
Silphium perfoliatum L.	Asteraceae
Sorghum halepense (L.) Pers.	Poaceae
Spartina alterniflora Loisel.	Poaceae
Spartina argentinensis Pers.	Poaceae
Spartina densiflora Brongn.	Poaceae
Spartina maritima (Curtis) Fernald	Poaceae
Spartina pectinata Bosc ex Link	Poaceae
Stenotaphrum secundatum (Walt.) Kuntze	Poaceae
Thalia geniculata L.	Marantaceae
Thrinax radiata Lodd. ex Schult. and Schult.	Arecaceae
Thysanolaena maxima Kuntze	Poaceae
Triglochin procerum L.	Juncaginaceae
Typha angustifolia L.	Typhaceae
Typha capensis Rohrb.	Typhaceae
Typha domingensis Pers.	Typhaceae
Typha latifolia L.	Typhaceae
Typha orientalis C. Presl	Typhaceae
Zizania caduciflora (Turcz. ex Trin.) Hand.-Mazz.	Poaceae
Zizaniopsis bonariensis (Balansa and Poitr.) Speg.	Poaceae

Table 1. Synopsis of vascular plants used for constructed greed wetlands with horizontal surface for wastewater treatment.

2.1. Phragmites

It is one of the most used genera. *Phragmites australis* (*Phragmites communis* Trin.) has been tested in Europe, Canada, Australia, many countries of Asia (except India and Nepal, where they use *Phragmites karka*) and many of Africa (except Central Africa, where it is used *Phragmites mauritianus*). In the USA and New Zealand, it has given good experimental results, but actually the utilisation of common reed has been limited because it is considered an invasive plant species [1, 14].

Seed germination of *P. australis* reaches germinative percentages up to 96–99% under the pretreatment of soaking the seeds with 0.1% KNO_3, rinsing them with distilled water before they are sown on layers of Whatman grade no. 1 filter paper (pH 7) in 90-mm Petri dishes. They must be moisten for up to 10 days and maintained into an incubator, with alternating diurnal regime of 12-h daylight at 25°C and 12 h of darkness at 15°C [15]. Fourteen hours/25°C and 10 h/20°C regime can also be applied [16]. Watering everyday with 10 ml of 9-mM sulphide solution increases significantly the germination speed [16]. Arbuscular mycorrhizal (AM) fungal inocula of *Funneliformis mosseae* accelerate seed germination of the species and, most important, enhance growth and development of its seedlings, performing as an efficient bio-accelerator, bio-fortifier and bio-enhancer [17]. Other proposals suggest using 1% agar as a germination medium and the following light and temperature conditions: 12-h/12-h photoperiod and 33/19°C (86% germination), 12-h/12-h photoperiod and 26/16°C (93% germination) and 8-h/16-h photoperiod and 35/20°C (95% germination [18]. Chemicals have a differential effect on *P. karka* seed germination. In complete darkness, as well as in 12-h light, 12-h dark photoperiod and different temperature regimes (10/20°C, 15/25°C, 20/30°C), seed germination is significantly promoted by thiourea (10 mM), nitrate (20 mM), proline (0.1 mM), betaine (0.1 mM), GA_3 (3 mM), kinetin (0.05 mM) or fusicoccin (5 μM) [19]. 5- and 10-mM ascorbic acid solutions have also given good results [20].

2.2. Typha

Cattails are very productive plants with maximum above-ground biomass values in constructed wetlands. Treating systems with *Typha angustifolia, Typha capensis, Typha domingensis, Typha latifolia* and *Typha orientalis* have been reported in the USA, Central and South America, Asia and several European countries [1]. In Spain, floating systems with these plants have been developed as an interesting innovative green technology, QuarQ Enterprise [2] (**Figure 1**).

Typha seeds and seed heads need to be cleaned in a seed cleaner before they are sown [21]. A strong jet of distilled water can be used for this purpose. Afterwards seeds will be settled in deionised water to select the most viable ones, which sink, whereas non-viable ones float [22]. An immersion for 24 h in slightly saline water with a concentration of up to 1% ClNa and pH 6.5–7.5 has also been proposed [23]. Specialised sources of seeds are recommended by the United States Department of Agriculture (USDA) [21], who indicates typha seeds germinate readily when they are planted in clean, moist seed bed and maintained for about 2 weeks in a greenhouse in pots 1 cm under the soil surface. Greenhouse temperature should be 37 ± 3°C.

Figure 1. QuarQ Enterprise Water Technologies in Villafranco, Badajoz (Spain).

T. angustifolia seeds can germinate with maximum success (100% germination) watering with distilled water, maintaining temperatures of 35/20°C and programming an 8-h/16-h photoperiod. Seed scarification does not seem to be worthwhile, because 58% of germination is obtained with this pretreatment and the above-mentioned germination conditions. When using 1% agar medium and germination conditions of 33/19°C, and a 12-h/12-h photoperiod, germination can reach 85% [18]. *T. domingensis* seeds do not germinate without light [24]. Germination up to 100% is obtained [24, 25] using an environmental chamber or an outdoor shade house covered by nylon netting (light: 10,764–32,292 lx). Petri dishes with paper towels or filter paper as substrate [25] and oscillating temperatures of 32/26°C or 29/21°C for 15 days [25]. A constant temperature of 30°C and a 12-h/12-h [23] (light: 5,000 lx) or 14-h/10-h [24] light/dark photoperiod can also be used. In the latter case, 5-mm layer of water-saturated *Sphagnum* peat has been tested with good results (85±13% germinative percentage). pH peat is adjusted to 7.0 by adding calcium carbonate (7.5 g $CaCO_3$ per pet's litre) [24]. To remove possible germination inhibitors and to retard the growth of microorganisms, seeds must be previously washed with running water and sodium hypochlorite solution (10%), commercial bleach [24], and furthermore immersed in a germination activator solution with ammonium and phosphate, pH 6.5–7 [23]. Pretreatments with 0.1% $KMnO_4$ are also recommended [15]. Another medium quite useful to be used is 1% agar. This combined with the 8-h/16-h photoperiod can give very interesting germination results: 90–100% germination maintaining constant temperatures of 20–25°C during the germination period and 80–90% if temperature is elevated to 30°C [18]. Alternating temperatures of 35/20°C with the same photoperiod and germinative substrate produce 100% germination as well. Changing the photoperiod to a 12-h/12-h rhythm can give good results (86% germination) at a constant temperature of 31°C, and it is excellent (98–100% germination) when a 33/19°C alternating programme is applied [18]. *T. latifolia* seed germination success depends on light, pH and alternating temperatures. It is inhibited by total obscurity and limited at low levels of pH [22] but scarcely affected by anoxic conditions [26]. Germination rates can reach 84% on mesic peat (pH 4.3), 22°C and a 16-h/6-h (day/night) photoperiod [22] and maximum rates when using 20/30°C with 12 h/12 h photo-thermoperiod [27]. Excellent results (97% germination) can be obtained in similar conditions (19/33°C, 12 h/12 h) on 1% agar germination medium [18]. Pre-sowing treatments are also recommended, moist in high humidity over water for 1 day at 20°C, and then removing and chipping the covering structure [18]. After that, seeds can be sown in 1% agar germination medium and 8-h/16-h photoperiod, obtaining 78% germination at a constant temperature of 30°C and 88–84% germination using alternating temperatures of 30/15°C and 30/20°C, respectively [18]. In pre-sowing as above-mentioned and fitting the latter photo-thermoperiod (30/20°C, 8 h/16 h), 76% germination has been reported for a medium containing 1% agar + 101-mg/l potassium nitrate (KNO_3) [18]. *T. orientalis* seeds germinate 100% in 1% agar and 35/20°C, 8-h/16-h photo-thermoperiod [18] (**Figure 2**).

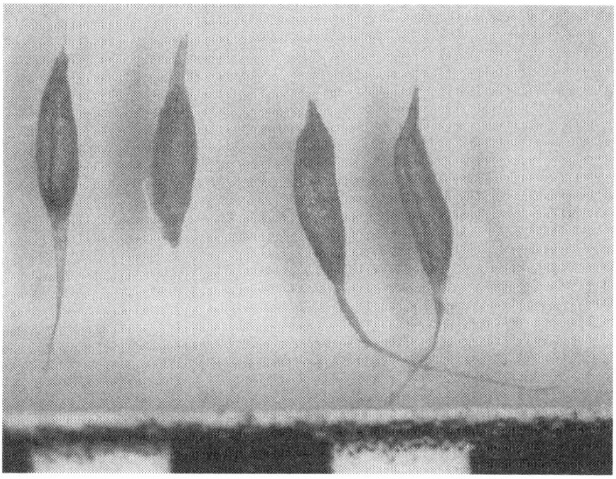

Figure 2. *Typha latifolia* seeds.

2.3. Scirpus

Different species of *Scirpus* (*Schoenoplectus*) have been tested in the USA, China, Australia and New Zealand (*Scirpus acutus, Scirpus americanus, Scirpus californicus, Scirpus cyperinus, Scirpus fluviatilis, Scirpus grossus, Scirpus lacustris, Scirpus maritimus, Scirpus pungens, Scirpus sylvaticus, Scirpus tabernaemontani, Scirpus validus*) [1, 5, 6]. Mexico and France have experimented with *S. validus* and *S. maritimus*, respectively. Sewages wastewaters have been frequently treated with these plants [1]. *S. grossus* has been tested for heavy metals content in Malaysia [3].

S. americanus seeds need light to naturally break dormancy and germinate [28] and/or to be subjected to cold stratification (3–6°C) for 30–180 days [18]. Germination conditions of 30–32°C, 12 h/12 h or 35/20°C are proposed [18]. When germination processes are taking place at greenhouses, it has been suggested to sow seeds in a cold frame pot standing in three centimetres of water. The seeds germinate quickly. When they are large enough to handle, they must be planted into their permanent positions in early summer [21]. Alternatively, a container (pot or flat) can be used. It should be watered from the bottom as necessary, and it should not be covered after sowing, although a light dusting of soil can be applied. If grown in outdoor beds, seeds are sown on level soil and covered with a single layer of burlap or cotton sheet. Soil dry must be avoided, shading with a window screen set 30 cm [28]. Procedures to maximise seed germination of *Scirpus acutus* have been studied in the laboratory [29]. Pregermination conditions included scarification and stratification at 4±1°C for 84 days [18, 29] while submerged in water [29]. Seeds were placed in night/day temperature regimes of 10/25°C under a 14-hour photoperiod (≈200 μmol m^{-2} s^{-1} photosynthetic photon flux density), and up to 97.5% germination was achieved [29]. At the greenhouse, similar requirements as for *S. americanus* were reported [28]. To germinate seeds of *S. californicus* in greenhouse, they must be introduced in greenhouse in 2.5 × 2.5 × 5 cm pots, 0.5 cm under the soil surface. Soil surface needs to be moisted and maintained at 35–40°C. Seeds begin to germinate after a couple of weeks. Plants are ready in 100–120 days to come out as plugs [21]. *S. cyperinus* seeds should be imbibed on agar 1% for 20 weeks at 5°C as a pre-sowing treatment that has been suggested to have best results [18]. At greenhouses, seeds should be sown in a cold frame as soon as they are ripe in a pot standing in three centimetres of water, and they will germinate easily [21]. A loam, peat and sand wet substrate can also be used [28]. An 8-h/16-h photoperiod is a good option to obtain germinative success; using 1% agar as medium, 100%, 98%, 96% and 90% germination can be reached setting the following alternating temperatures: 35/20°C, 30/15°C or 20/5°C, 25/10°C and 40/15°C [18]. Using 1% agar + 250-mg/l gibberellic acid (GA3), with the same photoperiod, 100% germination is got at alternating temperatures 20/5°C and 93% at 25/10°C [18].

S. fluviatilis seeds germinate after a period of moist, cold stratification. They need to be mixed with equal amounts or more of damp sand, vermiculite or other sterile media and introduced in a plastic bag and maintained at 0–3°C for 3 months. Some seeds may sprout in the storage bag if moist stratified too long. If sprouting occurs, seeds must be immediately planted. Another method of breaking dormancy for this species at temperate climates or latitudes is to sow seeds outdoors in the fall so they may overwinter [28]. *S. lacustris* germination is reported to occur when using a pre-sowing treatment of cold stratification for 80 days and a later germination phase with alternating temperatures of 30/5°C [18]. Presoaking seeds in sodium hypochlorite and performing cold stratification under light conditions are presented for consideration as a good tool to improve germination results with this species [30]. *S. maritimus* cold stratification for 80 days is profitable before planting seeds to germinate under 30/5°C [18]. Stratification period can be reduced to 28 days if seeds are presoaked in sodium hypochlorite and kept under natural light conditions [30]. Mechanical pretreatment of the seeds to evade physical dormancy has also been stated [31]. *S. pungens*, opposite to the abovementioned Scirpus species, does not seem to respond so positively to the mechanical scarification (by squeezing with tweezers) or the stratification in cold water, although further studies on this taxon must be performed [32]. *S. sylvaticus* germination works very well (80% germination), imbibing for 56 days the seeds on 1% agar at 6°C and maintaining them afterwards on a 1% agar medium and a thermo-photoperiod of 33/19°C, 12 h/12 h [18]. Even better results (89% germination) can be obtained replacing the pretreatment by adding to the 1% agar, gibberellic acid (GA3) 250 mg/l [18]. *S. tabernaemontani* pre-sowing treatments have been proposed: (a) cold stratification for 80 days [18], (b) imbibing the seeds for 56 days in 1% agar at 5°C [18] and (c) introducing them for 42 days into a sodium hypochlorite solution under low temperatures and natural light conditions [30]. During the germination period, fluctuating temperatures are recommendable [30], 30/5°C [17] and 35/20°C [30]. In the latter conditions, the use of 1% agar as medium and a photoperiod of 8 h/16 h assert germinative percentages up to 84% germination [18]. *S. validus* seeds germinate after a period of [28] moist cold stratification of 180 days [18]. They need light to naturally break dormancy [27] and germination thermic conditions of 30–32°C [18]. They can easily be sown in flat which will be watered from the bottom as necessary. A light dusting of soil can be applied slightly, covering the seeds. If they are grown in outdoor beds, they can be sown on level soil, covering it with a single layer of burlap or cotton sheet, shading it with a window screen (set 30 cm) [28].

2.4. Cyperus

Several species from *Cyperus* genus (*Cyperus alterniflorus, Cyperus articulatus, Cyperus dubius, Cyperus esculentus, Cyperus flabelliformis, Cyperus grandis, Cyperus immensus, Cyperus involucratus, Cyperus isocladus, Cyperus malaccensis*) have been tested in Asia (China, Thailand), America (Nicaragua, Brazil), Africa (Kenya) and New Zealand among other countries [1]. Some species, such as *C. alterniflorus*, have recently been used in a pilot scale in Italy [7]. Most of them are not sufficiently studied in terms of the seed technology.

C. alterniflorus germinates 100% utilising 1% agar medium and alternating temperatures of 25/10°C or 35/20°C and a photoperiod of 8 h/16 h [18]. *C. articulatus* seed germination is easy going if a photoperiod of 8 h/16 h is fixed and a basic 1% agar medium is used for the process. No pretreatments are required. Maximum results are obtained at different alternating temperatures: 100% at 40/25°C, 98–100% at 35/20°C, 96% at 30/15°C and 88% at 25/10°C. By adding gibberellic acid (GA3) 250 mg/l to the agar medium, we can obtain 100% germination at 30/15°C, 98% at 40/25°C, 96% at 25/10°C and 92% at 35/20°C [18]. *C. dubius* can be successfully germinated (76%–85%) sowing the seeds in 1% agar medium, an 8-h/16-h photoperiod and alternating temperatures of 35/20°C and 30/15°C, respectively [18]. *C. esculentus* germination of the seeds is significantly influenced by both light and temperature. It is highest at 35°C, and poor germination was observed at other temperatures (27 and 45°C). The plant growth regulators enhance the seed germination and radical length to a different degree [33]. *C. flabelliformis* germination percentage can reach 100% when using 1% agar as medium and temperature and light conditions of 35/20°C, 8 h/16 h. Cold-wet stratifications as pretreatments should not be planned because they reduce germination to 80% [18]. *C. involucratus* can be propagated by seeds in temperate latitudes at 18 to 21°C in spring in constantly moist seed compost [34]. *C. malaccensis* germinative process is mediated by arbuscular mycorrhizal colonisation, which is influenced at the same time by pH and moisture [35].

2.5. Carex

Some sedges (*Carex acutiformis, Carex gracilis, Carex lacustris*) have also been employed as phytoremediators in temperate regions (East Europe) [1], with recent applications in the east of Europe countries [12]. Very few species have been studied to date, and little is known about the technologies to improve the germination of their seeds.

It is reported that *C. acutiformis* seeds should be undergone to a pre-sowing treatment consisting in maintaining them for 2–6 months at 4°C. After that, they are placed of moist filter paper at germination conditions of 22/10°C, and an up to 65% germination is expected [18]. *C. gracilis* cold-wet stratification is suggested for a successful germination process (more than 80% germination). For that, seeds can be wetted with distilled water and introduced in 100-cc plastic vials, which will be sealed, wrapped in aluminium foil and stored at 4°C in a refrigerator for 6 months. After that, seeds must be placed at an incubator with alternating temperatures of 22/10°C [36]. *C. lacustris* has conditionally dormant seeds which require cold stratification [37]. The species is very sensitive to storage conditions [38]. Germination temperature regimen of 27/15°C ensures (at least 50%) discrete results [37]. Better results can be obtained by using seeds produced earlier in the same growing season. They must be situated into highly humid soils. Suitable soil amendments should be added to adjust the proper organic matter content to levels found in natural sedge meadows [38].

2.6. Other Cyperaceae

Eleocharis sphacelata, Kyllinga erecta and *Lepironia articulata* have been locally used in Australia, Tanzania and China [1]. For *Eleocharis sphacelata*, pretreatment, timing and water depth available for germination play an important part on the sexual multiplication of this plant, but there is still much to know about the specific requirements [39]. *Kyllinga erecta* seeds fail to germinate in darkness [40]. 85% germination has been recorded by the application of a pre-sowing treatment (imbibing on 1% agar for 12 weeks at 5°C), a germination medium of 1% agar and germination conditions of 35/20°C and photoperiod 8 h/16 h [18]. For *Lepironia articulata* a 50% germination was achieved in seeds buried 4 months, exhumed and incubated in anoxic/dark conditions [41].

2.7. Bambusoid-like species

Some bambusoids as *Arundo donax, Brachiaria mutica, Coix lacryma-jobi, Gynerium sagittatum, Pennisetum purpureum, Phylidrum lanuginosum, Thysanolaena maxima, Zizania caduciflora* and *Zizaniopsis bonariensis* have been utilised in Morocco, El Salvador, Costa Rica, Jamaica, Central

America, Australia, Mozambique, China and Brazil [1, 9]. But the multiplication by seeds of most of these species is barely developed.

Arundo donax seed germination is much better known than the rest of the bambusoid-like species. Maximum results (100% germination) are got using 1% agar medium and thermic and light conditions of 35/20°C, 8 h/16 h. Several pre-sowing treatments (cold shock −80°C, smoke and heat shock) have demonstrated not to be recommendable, especially if combined with constant temperature regimes of 20°C [18]. On the other hand, hydro-seeding has emerged as a reliable way to be employed. It consists in mixing selected seeds with colloidal substances in an aqueous solution usually along with mulch of various origins and fertilisers. Commercial names and composition are the following: Provide Verde (complete mixture of soil microorganisms), *Penicillium* sp, algae, polysaccharides, mulch (1 L ha^{-1}), vegetal glue, organic fertiliser (4.4% N, 2.2% P_2O_5, 1.1% K_2O, 2.1% Mg, 2.7% Ca), Envitotal-1 (total nitrogen 1.7% , organic carbon 11%, zeolites 38%, vegetal mulch mixture of cellulose 33% and vegetal glue), Envitotal-2 (total nitrogen 1.7% , organic carbon 11%, zeolites 38%, vegetal mulch mixture of cellulose 33%, soil N-fixing microorganisms 2.6%, vegetal glue), and Cellugrun (cellulose fibre mulch of 80% cellulose content, pH 7.5, bulk density 20–35 g L^{1-}, average fibre length 1400 μm, average fibre thickness 45 μm) and an adhesive hydrocolloid compound as soil control [42]. For an excellent (100%) seed germination result in the case of *Coix lacryma-jobi*, they need to be placed in Petri dishes (11 cm in diameter) containing silica sand and 10-mL distilled water into a climate-controlled incubator at 25°C, for a 10 days period, with a light (10 h) and dark (14 h) cycle [43]. Alternating temperatures 35/20°C are not recommended for this species [18].

Pearl millet (*Pennisetum glaucum*) seeds need to be washed previously with liquid soap solution and bavistin (fungicide) for 3 min and rinsed twice with deionised water [44]. Then they must be placed in Petri dishes (90 mm in diameter), containing one moisted piece of filter paper Whatman no. 1, and after that they have to be covered and maintained on a lab bench at room temperature (30 ± 5°C) [44]. An incubator programme of 30/19°C 16 h/8 h has also been successfully used [45]. Better results (100% germination) can be offered if the germination medium is 1% agar and the conditions are constant temperatures of 21°C or 26°C and photoperiod 12 h/12 h or 20°C and photoperiod 8 h/16 h [18]. With the same medium, 80% germination is obtained fixing 33/19°C, 12 h/12 h, and 75% germination with 25°C, 8 h/16 h [18]. Maximum success (100%) is aimed by removing the seed coat and performing the germinative process at 25°C and an alternating light regime of 8 h/16 h [18]. Finally, in selected lines of this species (`MS 841A', `MS 841B' and `D 23'), some pretreatments have been investigated to ensure at least 75% germination. Thus storage containers (cloth bags, polylined cloth bags), storage environments (ambient, controlled) and seed dressings (carbendazim, captan, thiram, *Trichoderma viride* and *Pseudomonas fluorescens*) were tested. The germination of seeds stored under controlled conditions (82.25%) (temperature 20°C and relative humidity 40%) and in polylined bags (76.62%) was significantly higher than seeds stored under ambient condition (66.43%) and in cloth bags (72.05%). Treatment with bioagent *Trichoderma viride* gave a germination percentage of 77.37%, *Pseudomonas fluorescens* treatment 59.58% and untreated control 74.27%. The incidence of seed mycoflora was 32.39, 28.26 and 29.14% in seeds of 'MS 841A', 'MS 841B' and 'D 23', respectively. Seeds stored under controlled conditions germinated 35.31%, under ambient condition (24.55%) and maintained in cloth bags 31.21% and in polylined cloth bags 28.66%. The incidence of contamination with seed mycoflora was 40.24% in the untreated control, 6.90% in thiram treatment, 11.65% in captan treatment, 34.07% in *Trichoderma viride* treatment, 39.46% in carbendazim treatment and 47.28% *Pseudomonas fluorescens* treatment [46].

2.8. Non-cane-like Poaceae

In this group we have included species as *Echinochloa polystachya, Festuca arundinacea, Glyceria maxima, Panicum maximum, Panicum repens, Paspalum distichum, Phalaris arundinacea, Sorghum halepense, Spartina alterniflora, Spartina argentinensis, Spartina densiflora, Spartina maritima, Spartina pectinata* and *Stenotaphrum secundatum*, which are reported to be employed in Ecuador, Alabama, Kentucky and Washington (USA), New Zealand, Germany, Czech Republic, Jordan, Italy and Portugal [1]. Although there is some information on the seed technology of these taxa, many of them still require further development of proper methodologies adapted to each species.

Some interesting improvements have been made with *Festuca arundinacea*, for example. To ameliorate seed germination performance of it, hydropriming can be advised. Good results can be obtained by maintaining them in a germinator at 15 to 25°C for a period of 8 h of darkness and 16 h of light with a light intensity of 38 µmol m^{-2} s^{-1} provided by cool-white fluorescent lamps [47]. *Glyceria maxima* seed technology for germination has been more profusely tested. 100% germination has been obtained by sowing in 1% agar medium and maintaining one of the following regimes: 20°C, 8 h/16 h; 15°C, 8 h/16 h; 25/10°C, 8 h/16 h and 23/9°C, 12 h/12 h [18]. Using the same medium 95–96% germination was obtained when employing 21°C, 12 h/12 h, and 25°C, 8 h/16 h, respectively, and 85% germination if the conditions were 21/11°C, 12 h/12 h [18]. The latter result has been improved (94% germination) repeating the conditions but pretreating the seeds by imbibition on agar 1% for 4 weeks at 2°C [18]. Several tests adding 101-mg/l potassium nitrate to the agar medium did not give extremely advisable results, neither imbibing the seeds on agar 1% agar for 8 weeks at 6°C [18]. Opposite to this, an imbibition on 1% agar for 1 day at 20°C and then for 2 days at 4°C (germination conditions of 20/15°C, 12 h/12 h) has been reasonably recommended for it ensures 90% germination [18]. This species can also be germinated on damp filter paper into a Petri dish covered with a colourless plastic to allow light penetration and to prevent rapid moisture loss. They will be introduced in a grow chamber with a temperature range of 21–23°C with a 12-h/12-h (light/dark) cycle. If cold stratification at 4°C for 8 weeks has been implemented, more than 80% germination is reported [48].

For *Panicum maximum* recommended method for best (100%) seed germination is to sow them in 1% agar medium and introduce them in a room chamber at a constant temperature of 21°C and 12 h/12 h or a combined cycle 35/20°C, 8 h/16 h [18]. In the latter case, seeds must previously be immersed in 10% Domestos solution for 5 min [18]. Other scarifying procedures (shallow incisions, mechanical removing) decrease germination in this species [49] from percentages to 92%, 85%, 78%, 72% or even 40% [18]. If the available light cycle is 12 h/12 h, the former conditions should be maintained; increasing temperature to 26°C lightly decreases the percentage of germination to 90%, and changing to alternating 33/19°C reduces it to 75% or 80% (in the particular case that a solution of 101-mg/l potassium nitrate has been added to the agar medium) [18]. *Panicum repens* germinates 80% if seed coats are removed and seeds are germinated on 1% agar medium and photothermic conditions are 35/20°C, 8 h/16 h [18]. Some authors have pointed out [50] that *Phalaris arundinacea* germination does not occur in the dark [50] and that this species is photoperiod insensitive in the range of 12–16 hours. Upon them, the highest germination percentages (up to 80) can be obtained under white light and red light (11.0 μmol s^{-1} m^{-2}) and up to 40 with high red: far-red ratios [50]. On the other hand, 96–98% germination has been reported [18], scarifying the seeds and sowing them in 1% agar at germination conditions of 23/9°C, 12 h/12 h [18], and 90% at an 8-h/16-h photoperiod and 20°C [18]. Without removing the coat, 90% germination can be reached on 1% agar medium, fixing the incubator conditions at 25/10°C, 8 h/16 h [18], and 80% by adding to the agar a solution of 101-mg/l potassium nitrate and choosing the cycle 23/9°C, 12 h/12 h [18].

Sorghum halepense germinates perfectly (100% germination) on 1% agar medium at a constant temperature of 26°C and a day/light rhythm of 12 h/12 h or at an alternating 35/20°C, 8 h/16 h, if seed coat has previously been removed [18]. *S. halepense* has neutral photoblastic seeds and presents mechanical-type dormancy [51]. Different sorts of scarification have been proposed: sandpaper [51], scalpel, pericarp excision from along proximal to distal ridge above embryo and previous imbibition on 1% agar for 18 weeks at 20°C, or at 25/10°C, or for 8 weeks at 6°C [18]. In the 12-h/12-h light regime, alternating temperatures (23/9°C) reduced germination to 80%, so it seems better to maintain it constant at the cited value (26°C) [18]. If an 8-h/16-h photoperiod is of our convenience, we should consider that the thermic alternation of 25/10°C lowers germination to 89%, even for scarified seeds [18]. As well fixing temperatures to 20°C or 30°C brings germination percentages down to 85%, despite the scarifying [18]. Adding 101-mg/l potassium nitrate to the agar medium has not improved the germination results in many tests made in these conditions. It is not a best choice, for it may get down germination even to 78% [18]. *Spartina alterniflora* germination of the seeds is not affected by light or dark, and the optimal temperature 16/26°C (night/day) gives a germination rate >90% [52]. Salinity does not reduce seed germination of this halophilous species [53] when it does not exceed to 450-mM NaCl [54]. Improved smooth cordgrass cultivars of easy germination, 'St. Bernard' (LA12-101) (Reg. No. CV-268, PI 665014), 'Las Palomas' (LA12-102) (Reg. No. CV-269, PI 665015) and 'Lafourche' (LA12-103) (Reg. No. CV-270, PI 665016), are available for implementation [55]. Good results have been obtained in climate chambers at 25°C constant temperature or alternating 20/30°C, much better [56]. In the case of the close taxa *S. pectinata*, 91% germination has been reported scarifying the seed before sowing and doing it on 1% agar at 25/10°C, 8-h/16-h conditions [18].

2.9 Juncoid species

We include here plants belonging to Juncaceae (*Juncus effusus, Juncus inflexus, Juncus subsecundus*), Juncaginaceae (*Triglochin procerum*) and *Hemerocallis fulva* (Xanthorrhoeaceae) that have been utilised for wastewater treatment in Australia, Kentucky (USA), Portugal, Germany, Canada, Slovenia and Spain [1, 10].

Juncus effusus seeds need light, moisture and heat for germination. To decrease the time the seed takes to sprout, two methods have been proposed: soaking the seeds [21] and imbibing them on 1% agar for 8 weeks at 2°C [18]. To grow seeds in greenhouses, they will be placed on soil surface, pressing lightly to assure good soil contact, not covering them and keeping the soil moist. Greenhouse should be kept hot (32–38°C). Seeds begin to germinate in approximately 1 week. Soil moisture has to be maintained until plants are to be transplanted [21]. In incubators, to ensure 97% germination, the medium suggested is 1% agar + 250-mg/l gibberellic acid (GA3). Germination conditions are alternating temperatures and photoperiod of 23/9°C, 12 h/12 h. If no scarification is made and no GAE is used, germination may be reduced to 91% (35/20°C), 8 h/16 h, or even 88% (30°C, 8/16) [18]. *J. inflexus* seed germination is a challenge, and the best published results about their seed germination reach just 96%. Basic 1% agar has been repeatedly used as germination medium. Germination cycles of 23/9°C, 12 h/12 h, and 35/20°C, 8 h/16 h, let to reach 95%. With the same photoperiod and medium, changes in the thermic programme decrease the germinative success to 88% (25/15°C, 8 h/16 h), 86% (25°C, 8 h/16 h) and 70% (25/15°C, 8 h/16 h), respectively. Moving the light plan to a 12-h/12-h cycle and the temperatures to 23/9°C has reduced the germination to 50% [18]. Pretreatments have also been tested. Chipping with scalpel has given good results (94–96% germination) when combined with adequate photothermic programmes: 26°C, 12 h/12 h. Other proposals have been to water the seeds in high humidity over water for 1 day at 20°C and then imbibing on 1% agar for 8 weeks at 5°C. In this case 85% germination has been got by applying the photothermic routine 20/10°C, 8 h/16 h, and 95% by just changing the photoperiod to 12 h/12 h. It does not seem so recommendable to use another pre-sowing treatments published, vg. 1% on agar imbibition for 8 weeks at 5°C at 20/10°C, 8 h/16 h, or agar imbibition during 6 weeks at 20/10°C, 8 h/16 h, using 25°C, 8 h/16 h, and 25/10°C, 8 h/16 h, Respectively, bacause they let to reach 88% and 75%. On this occasion, even using GA3 gibberellic acid (250 mg/l) as a complement of the agar medium [18]. Finally to germinate *J. subsecundus* seeds, 1% agar medium is suggested, maintaining 20°C, 8 h/16 h, as to reach to 75% germination [18].

2.10. Flowering ornamental monocots

Decorative perennial herbs are easily available and can grow well under local climatic conditions. Among monocotyledons, many species have been used, especially for on-site treatments where aesthetic or look of the place is an important factor. Some of those species are *Acorus calamus, Hymenocallis littoralis, Colocasia esculenta, Canna glauca, Canna indica, Canna x generalis, Heliconia psittacorum, Heliconia rostrata, Iris pseudacorus, Iris tectorum, Iris versicolor* and *Thalia geniculata*. There are reports of their use in Tanzania, Ohio, Kentucky (USA), Mozambique, Brazil, Mexico, Colombia, El Salvador, Czech Republic, Estonia and Portugal [1, 7, 8].

Acorus calamus is an emergent macrophyte with great potential for use in wetland restoration in North America. Seed germination occurs only in full light at 15/25°C or 20/35°C and seeds fully submerged [7]. In temperate climates, they also can be planted in a greenhouse during the fall or winter. A 5-cm deep tray filled with an organic-soil mix can be used. Seeds should be scattered sparsely on the surface and pressed firmly into the soil, burying them no further than 0.3-cm deep. Soil needs to be moist to saturation. Seed does not require stratification and germinates in less than 2 weeks. When plants get enough size, they must be transplanted [21]. *Colocasia esculenta* germ plasm can be conserved as seed for at least 2 years at constant 5°C and −20°C when seed moisture content is reduced to 10–12% and at ambiental room temperature (21.5–34.4°C, mean 27.2°C) when seed moisture content is reduced to 7.3% [57]. *Canna indica* is an herbaceous species with ornamental and medicinal value, having seeds with a hard seed coat. Seed germination is good at 10–40°C, being the optimal temperature range between 13.84 and 34.41°C, determined by the enthalpy of activation [58]. To open the imbibition lid, a raised incubation temperature of 50°C during 24 hours in wet surroundings is enough. This hydration induces germination. As a result, the integumentary part of the seed coat softens, making it possible for the germinal root to emerge from the seed. This water-regulating mechanism, combining an impermeable palisade layer and imbibition lid, is a unique feature of the Cannaceae. The seed coat is mainly of chalazal origin, and the main mechanical layer is formed by the exotesta. Other families of the Zingiberales, in contrast, open by an operculum formed by all seed coat layers. Moreover, the seed coat in those families is of integumentary origin, and the main mechanical layer is formed by the exo- and/or endotesta [59]. *Heliconia rostrata* seed germination is well known as difficult and problematic, because the embryo is not yet well differenced when the seed matures, and it has a hard testa which does not allow water to get into it. It can take from 3 months to 3 years [60]. *Iris versicolor* seeds germinate 58% in the greenhouse using cold stratification; this is storing seeds in wet paper towels at 4–5°C for 3–4 weeks [61]. *Iris pseudacorus* germinates 80–86% in an incubator at 12-h/12-h light regime and 30/20°C alternating temperatures or constant temperature of 26°C. In the latter case, seeds must be pretreated by immersion in 10% Domestos solution for 5 min and then imbibed on 1% agar for 8 weeks at 6°C, and then the seed is scarified (covering structure removed and seed coat chipped). After that, a solution of 250-mg/l gibberellic acid (GA3) must be added to the 1% agar medium.

2.11. Megaforbics and other groups

We have included here species as *Asclepias incarnata, Hibiscus moscheutos, Filipendula ulmaria, Liatris pycnostachya, Lobelia cardinalis, Lythrum salicaria, Mentha spicata, Rudbeckia hirta* and *Silphium perfoliatum* [1] that have been tested in Ohio, Kentucky and Minnesota (USA).

To germinate *Asclepias incarnate* seeds, they must be placed into plastic bags filled with moist perlite or vermiculite and stored for 4–12 weeks in a cold (1–3°C) place. Good germination results have been reported without stratification by soaking the seed twice in 87°C for 12 hours. Germination trays can also be used, filling the cells with a commercial seedling mixture or a mix of *Sphagnum* peat moss and vermiculite and moistening them very well. Seeds should be gently pressed into the soil, three seeds per cell, and covered with a very thin layer of soil. It is recommended to keep the soil moist during germination by spraying or misting. Ambient temperatures should remain between 18 and 23°C. This species requires light for germination [21]. *Filipendula ulmaria* is successfully germinated (95%) using 1% agar as germination substrate and incubator conditions of 25/10°C, 8 h/16 h, or 23/9°C, 12 h/12 h

[18]. *Liatris pycnostachya* germinates in 1% agar medium with maximum percentages at a light-alternating cycle of 8 h/16 h, maintaining a constant temperature of 15°C and 93% if temperature is elevated to 20% [18]. Stratification is needed to obtain efficient germination in this genus. Benzyladenine and thiourea are considered good agents for breaking dormancy. Fixing the germination chamber at 20°C and stratification at 4°C for 10 weeks produced 98% germination. Similar good results were obtained with the application of the following pretreatments: aqueous solutions of benzyladenine at 10 or 100 mg/l are applied to blotter paper; dry seeds treated for 3 minutes in benzyladenine at 0 to 1126 mg/l and dissolved in Acetone; a 3-minute acetone permeation of seeds with benzyladenine at 225 or 1127 mg/l; or seeds immersed in thiourea at 0.76 or 7.61 mg/l for 24 hours. Nevertheless gibberellic acid GA3 at 1, 10 or 100 mg/l in H_2O did not show significant efficiency [62]. *Lobelia cardinalis* seeds will germinate without cold stratification, but they need light. For this reason they can be sown in a flat with a damp fine grade peat light mix. If we keep the flats moist and under lights or in a greenhouse, they should green up in a few weeks. Afterwards they can be transplanted in 4–6 weeks into individual pots [21]. At an incubator, 92% germination is reported for the germination medium 1% agar and the germination conditions 15°C, 8 h/16 h [18]. *Lythrum salicaria* pre-sowing treatments have been fully described. Imbibition on 1% agar for 8 weeks at 6°C, 5°C and 2°C and further cultivation on 1% agar medium have given excellent results. 100% germination (21/11°C, 12 h/12 h, or 20/10°C, 8 h/16 h) and 95% germination (21/11°C, 12 h/12 h). Similar results have been reported even without that pre-sowing treatments but fixing the specific photothermic programmes of 35/20°C, 8 h/16 h (100%), or 20°C, 8 h/16 h (95%). Lower levels (84%) are obtained at 25/10°C, 8 h/16 h. The addition of 250-mg/l gibberellic acid (GA3) to the agar germination medium does not seem to have a clearly positive effect, although 92% germination is reached when using 12 h/12 h and 21/11°C and 88% if applying 8 h/16 h and 20/10°C [18]. *Mentha* spp. cultivation has been studied from different perspectives. Recent studies have explored the influence of high-frequency pulsatile electromagnetic fields and ultrasound pulsatile fields to improve mint seed germination. Further studies are needed, but there are promising first results [63]. In *Mentha spicata*, optimal temperature for seed germination is 30°C. It could be affected under different light qualities and full darkness conditions. White light is the most conductive to seed germination, but full darkness is the least conductive. After soaking treatment with GA 3 0.25% KNO_3, seed germination percentage is significantly promoted [64]. *Rudbeckia hirta* can be easily multiplied by seeds cultivating them on 1% agar medium in an incubator at 21°C, 12/12 (100% germination) or at 10°C, 15°C and 25°C and 8-h/16-h light cycle (84%, 89%, 89%, respectively) [18]. Finally *Silphium perfoliatum* seeds germinate 79% in 1% agar, alternating temperatures of 25/10°C and photoperiod 8 h/16 h [18].

3. Final comments

Seed technology for multiplication helophytes needs further experimental work. Significant efforts have been done, but it is a challenge to have a set of available information ready to be transferred to the productive sectors that are interested in this type of plants. Plants used in constructed wetlands with horizontal subsurface flow have a promising future as sustainable elements of the upcoming sewage equipment. For this reason improvements on the biotechnology of these species are a worthwhile researching line, where seed technology can play an important role.

Author details

Trinidad Ruiz-Téllez[1*], Francisco M. Vázquez-Pardo[2], José Blanco-Salas[2] and F. Javier Carbonell-Espín[3]

*Address all correspondence to: truiz@unex.es

1 University of Extremadura, Badajoz, Spain

2 La Orden-Cicytex Research Center, Badajoz, Spain

3 QuarQ Enterprise, Badajoz, Spain

References

[1] Vymazal J. Plants used in constructed wetlands with horizontal subsurface flow: a review. Hydrobiologia. 2011;674(1):133–56.

[2] Vasco Torrado V. aQuarQ System: OGESA bet on helophytes for water treatment in small settlements [Internet]. i-agua magazine 9. 2015 [cited 2016 Jan 13 23]. p. 76–9. Available from: http://www.iagua.es/noticias/espana/ogesa/16/01/20/14 sistema-aquarq-apuesta-ogesa-helofitas-depuracion-pequenas

[3] Tangahu BV, Abdullah SRS, Basri H, Idris M, Anuar N, Mukhlisin M. Phytoremediation of wastewater containing lead (Pb) in pilot reed bed using scirpus grossus. Int J Phytoremediation. 2013;15(7):663–76.

[4] Carpenter D, Boutin C, Allison JE, Parsons JL, Ellis DM. Uptake and effects of six rare earth elements (REEs) on selected native and crop species growing in contaminated soils. PLoS One. 2015;10(6):e0129936.

[5] Bhatia M, Goyal D. Analyzing remediation potential of wastewater through wetland plants: a review. Environ Prog Sustain Energy. 2014;33(1):9–27.

[6] Wang Y, Wang J, Zhao X, Song X, Gong J. The inhibition and adaptability of four wetland plant species to high concentration of ammonia wastewater and nitrogen removal efficiency in constructed wetlands. Bioresour Technol. 2016;202:198–205.

[7] Pai A, McCarthy BC. The influence of environmental factors on the germination of *Acorus calamus* seeds. Seed Sci Technol. 2010;38(2):390–8.

[8] Madera-Parra CA, Peña-Salamanca EJ, Peña MR, Rousseau DPL, Lens PNL. Phytoremediation of Landfill Leachate with Colocasia esculenta, Gynerum sagittatum and Heliconia psittacorum in Constructed Wetlands. Int J Phytoremediation. 2015;17(1–6): 16–24.

[9] Vanessa AE, Jonas FE, Maritane P, Edmar S de V, Marcos VMS, Kaian ACK, et al. The effect of swine raising wastewater in the development of millet (*Pennisetum glaucum* L.), soil and leachate. Afr J Agric Res. 2015;10(33):3206–15.

[10] Schultze-Nobre L, Wiessner A, Wang D, Bartsch C, Kappelmeyer U, Paschke H, et al. Removal of dimethylphenols from an artificial wastewater in a laboratory-scale wetland system planted with *Juncus effusus*. Ecol Eng. 2015;80:151–5.

[11] Zhang Y, Chen ZH, Zhang Z, Zhou XY, Tu Y. Ornamental plants in constructed wetlands. In: Ornamental Plants: Types, Cultivation and Nutrition. Nova Science Publishers, Inc.; 2011. p. 81–96. New York.

[12] Vergeles Y, Butenko N, Ishchenko A, Stolberg F, Hogland M, Hogland W. Formation and properties of sediments in constructed wetlands for treatment of domestic wastewater. Urban Water J. 2014;13(3):293–301.

[13] Albano Pérez E, Ruiz Téllez T, Sánchez Guzmán JM. Influence of physico-chemical parameters of the aquatic medium on germination of Eichhornia crassipes seeds. Plant Biol (Stuttg). 2011;13(4):643–8.

[14] Hazelton ELG, Mozdzer TJ, Burdick DM, Kettenring KM, Whigham DF. *Phragmites australis* management in the United States: 40 years of methods and outcomes. AoB Plants. 2014;6:1–19.

[15] Meng Huan, Wang Xuehong, Tong Shouzheng , LÜ Xianguo, Zhang Wenguang, Huo Lili, et al. Effects of pre-treatment on germination of Typha domingensis and *Phragmites australis*. Acta Ecol Sin. 2013;33(19):6142–6.

[16] Xiao Y, Sun J, Liu F, Xu T. Effects of salinity and sulphide on seed germination of three coastal plants. Flora – Morphol Distrib Funct Ecol Plants. 2016;218:86–91.

[17] Wu J, Ma F, Wang L, Yang J, Huang X, An G, et al. Seedling performance of *Phragmites australis* (Cav.) Trin ex. Steudel in the presence of arbuscular mycorrhizal fungi. J Appl Microbiol. 2014;116(6):1593–606.

[18] RBG. Seed Information Database: Royal Botanic Gardens, Kew [Internet]. 2016 [cited 2016 Jan 28]. Available from: http://data.kew.org/sid/

[19] Zehra A, Gul B, Ansari R, Alatar AA, Hegazy AK, Ajmal Khan M. Interactive effect of salt, light and temperature on seed germination and recovery of a halophytic grass—Phragmites karka. Pak J Bot. 2013;45(3):725–36.

[20] Zehra A, Shaikh F, Ansari R, Gul B, Khan MA. Effect of ascorbic acid on seed germination of three halophytic grass species under saline conditions. Grass Forage Sci. 201329;68(2):339–44.

[21] USDA. PLANTS Database | USDA PLANTS [Internet]. PLANT GUIDE. 2016 [cited 2016 Jan 27]. Available from: http://plants.usda.gov/java/

[22] Bourgeois B, Hugron S, Poulin M. Establishing a moss cover inhibits the germination of *Typha latifolia*, an invasive species, in restored peatlands. Aquat Bot. 2012;100:76–9.

[23] Semenova A. Method for producing aquatic plants for the purification of contaminated water. EP: PCT; WO2014131930 A1, 2014.

[24] Lorenzen B, Brix H, McKee KL, Mendelssohn IA, Miao S. Seed germination of two Everglades species, *Cladium jamaicense* and *Typha domingensis*. Aquat Bot. 2000;66(3): 169–80.

[25] Stewart H, Miao SL, Colbert M, Carraher CE. Seed germination of two cattail (Typha) species as a function of Everglades nutrient levels. Wetlands. 1997;17(1):116–22.

[26] Fraser LH, Mulac K, Moore FB-G. Germination of 14 freshwater wetland plants as affected by oxygen and light. Aquat Bot. 2014;114:29–34.

[27] Lombardi T, Fochetti T, Bertacchi A, Onnis A. Germination requirements in a population of *Typha latifolia*. Aquat Bot. 1997;56(1):1–10.

[28] Wildones. Germination Instructions Final Revision.pdf – Buscar con Google [Internet]. [cited 2016 Jan 28]. Available from: https://www.google.es/search?q=004cGerminationInstructionsFinalRevision.pdf&ie=utf-8&oe=utf-8&gws_rd=cr&ei=21uqVpXxJojSUYO1vUg#q=004c+Germination+Instructions+Final+Revision.pdf

[29] Thullen JS, Eberts DR. Effects of temperature, stratification, scarification, and seed origin on the germination of *Scirpus acutus* Muhl. seeds for use in constructed wetlands. Wetlands. 1995;15(3):298–304.

[30] Clevering OA. Germination and seedling emergence of *Scirpus lacustris* L. and *Scirpus maritimus* L. with special reference to the restoration of wetlands. Aquat Bot. 1995;50(1): 63–78.

[31] Brochet A-L, Guillemain M, Gauthier-Clerc M, Fritz H, Green AJ. Endozoochory of Mediterranean aquatic plant seeds by teal after a period of desiccation: determinants of seed survival and influence of retention time on germinability and viability. Aquat Bot. 2010;93(2):99–106.

[32] Giroux J-F, Bédard J. Seed production, germination rate, and seedling establishment of *Scirpus pungens* in tidal brackish marshes. Wetlands. 1995;15(3):290–7.

[33] Shen Y, Chen WJ, Lei XT, Shao HB, Tang MM, Li Y. Effects of plant growth regulators and temperature on seed germination of yellow nut-sedge (*Cyperus esculentus* L.). J Med Plants Res. 2011;5(31):6759–65.

[34] RHS. Royal Horticultural Society . Gardening [Internet]. 2016. [cited 2016 Jan 28]. Available from: https://www.rhs.org.uk/plants/details?plantid=2604

[35] Sridhar KR, Roy S, Sudheep NM. Assemblage and diversity of arbuscular mycorrhizal fungi in mangrove plant species of the southwest coast of India. In: Mangroves: Ecology, Biology and Taxonomy. Nova Science Publishers, Inc.; 2011. p. 261–78. New York.

[36] Schütz W, Rave G. The effect of cold stratification and light on the seed germination of temperate sedges (Carex) from various habitats and implications for regenerative strategies. Plant Ecol. 1999;144(2):215–30.

[37] Kettenring KM, Galatowitsch SM. Temperature requirements for dormancy break and seed germination vary greatly among 14 wetland Carex species. Aquat Bot. 2007;87(3): 209–20.

[38] Van Der Valk AG, Bremholm TL, Gordon E. The restoration of sedge meadows: seed viability, seed germination requirements, and seedling growth of Carex species. Wetlands. 1999;19: 756–64.

[39] Kapa MM. Ethnobotany, germination and growth of eleocharis sphacelata. The University of Waikato; 2009. New Zealand.

[40] Lowe DB, Whitwell T, McCarty LB, Bridges WC. *Kyllinga brevifolia, K. squamulata*, and *K. pumila* seed germination as influenced by temperature, light, and nitrate. Weed Sci. 1999;47(6):657–61.

[41] Lukacs GP, Weber E. Seed germination requirements of four emergent macrophytes in North Queensland, Australia. Australian Centro for Tropical Freshwater Research Report 97/21. 1997.

[42] Scordia D, Zanetti F, Varga SS, Alexopoulou E, Cavallaro V, Monti A, et al. New insights into the propagation methods of Switchgrass, Miscanthus and Giant Reed. BioEnergy Res. 2015;8(4):1480–91.

[43] LI J, JIN Z. Potential allelopathic effects of Mikania micrantha on the seed germination and seedling growth of Coix lacryma-jobi. Weed Biol Manag. 2010;10(3):194–201.

[44] Parveen A, Rao S. Effect of Nanosilver on Seed Germination and Seedling Growth in Pennisetum glaucum. J Clust Sci. 2014;26(3):693–701.

[45] Sobrinho FS, José da Silva Lédo F, Pereira A Vander, Silva e Oliveira J. Evaluation of potential propagation by seed of hexaploid elephant grass. Cienc e Agrotecnologia. 2008;32(3):974–7.

[46] Gupta A. Integrated management of seed health for prolonging seed storability in parental lines of pearl millet (*Pennisetum glaucum*). Indian J Agric Sci. 2007;77(4):212–9.

[47] Behtari B, Tilaki GAD. Changes in germination characteristics and seedling growth between storage and non-storage of primed tall fescue seed. Afr J Biotechnol. 2012;11(19):4317–22.

[48] Mugwedi LF, Goodall J, Witkowski ET, Byrne MJ. The role of reproduction in Glyceria maxima invasion. African J Range Forage Sci. 2014;32(1):59–66.

[49] Usberti R, Martins L. Sulphuric acid scarification effects on *Brachiaria brizantha, B. humidicola* and *Panicum maximum* seed dormancy release. Rev Bras Sementes. 2007;29(2):143–7.

[50] Lindig-Cisneros R, Zedler J. Effect of light on seed germination in *Phalaris arundinacea* L. (reed canary grass). Plant Ecol. 2001;155(1):75–8.

[51] Krenchinski FH, Albrecht AJP, Albrecht LP, Villetti HL, Orso G, Barroso AAM, et al. Germination and Dormancy in Seeds of Sorghum halepense and Sorghum arundinaceum. Planta Daninha. 2015;33(2):223–30.

[52] Yuan ZN, Shi FC. Ecological adaptability of alien species *Spartina alterniflora* Loisel seed germination. Chin J Ecol. 2009;28(12):2466–70.

[53] Elsey-Quirk T, Middleton BA, Proffitt CE. Seed flotation and germination of salt marsh plants: the effects of stratification, salinity, and/or inundation regime. Aquat Bot. 2009;91(1):40–6.

[54] Yuan Z, Shi F. Ecological adaptation strategies in alien species: effects of salinity, temperature and photoperiod on *Spartina alterniflora* Loisel. Seed germination. Pol J Ecol. 2009;57(4):677–84.

[55] Knott CA, Materne MD, Utomo H, Subudhi PK, Baisakh N, Harrison SA. Registration of "St. Bernard", "Las Palomas", and "Lafourche" smooth cordgrass cultivars. J Plant Regist. 2013;7(1):12.

[56] Zhu Z, Zhang L, Xiao D. Seed production of Spartina alterniflora and its response of germination to temperature at Chongming Dongtan, Shanghai. Shengtai Xuebao/ Acta Ecol Sin. 2011;31(6):1574–81.

[57] Rice T. Effect of seed storage under ambient and cold temperatures on germination of taro seed (Colocasia esculenta). SEED Sci Technol. 2007;35(3):674–87.

[58] Souza PRM, Tozzi HH, Takaki M. Temperature effect on seed germination in *Canna indica* L. (Cannaceae). Seed Sci Technol. 2011;39(1):243–7.

[59] Maas-Van De Kamer H, Maas PJM. The cannaceae of the world. Blumea J Plant Taxon Plant Geogr. 2008;53(2):247–318.

[60] Benítez-Domínguez L, Gómez-Merino FC, Trejo-Téllez LI, Robledo-Paz A. Anatomía, contenidos de ácido abscísico y nutrimentos y germinación de semillas de heliconia. Rev Fitotec Mex. 2011;34(3):189–96.

[61] Wees D. Stratification and priming may improve seed germination of purple cone flower, blue-flag iris and evening primrose. Acta Hortic. 2004; 629: 391–5.

[62] Parks CA, Boyle TH. Germination of *Liatris spicata* (L.) Willd. Seed is enhanced by stratification, benzyladenine, or thiourea but not gibberellic acid. HortScience. 2002;37(1):202–5.

[63] Singureanu V, Ungur R, Onac I, Kovacs MH, Moldovan G, Singureanu V. Automatic germination evaluation and qualitative analysis of essential oil of mentha × piperita L. under the influence of high frequency pulsatile electromagnetic and ultrasound pulsatile fields. Not Bot Horti Agrobot Cluj-Napoca. 2015;43(1):146–52.

[64] Liu JC, Gong WH, Xiang F, Yang JX, Fu ZR, Peng XM. Effects of different treatments on seed germination of Mentha rotundifolia. Chin Tradit Herb Drugs. 2012;43(5):1000–2.